Quick Reference
NEUROSCIENCE
for Rehabilitation
Professionals

The Essential Neurologic Principles

Underlying Rehabilitation Practice

Quick Reference
NEUROSCIENCE
for Rehabilitation
Professionals

The Essential Neurologic Principles
Underlying Rehabilitation Practice

Sharon A. Gutman, PhD, OTR

Assistant Professor, Division of Occupational Therapy
Long Island University
Brooklyn, NY

SLACK
INCORPORATED

an innovative information, education, and management company

6900 Grove Road • Thorofare, NJ 08086

Publisher: John H. Bond
Editorial Director: Amy E. Drummond
Design Editor: Lauren Biddle Plummer
Illustrator: Sharon Gutman, PhD, OTR

Gutman, Sharon A.
    Quick reference neuroscience for rehabilitation professionals: the essential neurologic principles underlying rehabilitatio practice/Sharon A. Gutman.
        p. cm.
    Includes bibliographical references and index.
    ISBN 1-55642-463-9 (alk. paper)
    1. Neurosciences. 2. Medical rehabilitation. I. Title.

RC343 .G88 2001
616.8--dc21

Library of Congress Control Number: 20001020093

Published by: SLACK Incorporated
                       6900 Grove Road
                       Thorofare, NJ 08086 USA
                       Telephone: 856-848-1000
                       Fax: 856-853-5991
                       www.slackbooks.com

Printed in the United States of America

DEDICATION

This book is dedicated in memory of my father Jesse, and to my mother Evelyn.
It is with much appreciation that I also dedicate this book to
Anne Hiller Scott, PhD, OTR, FAOTA,
Margery Szczespanski, MA, OTR,
and
Winifred Olivaria

Contents

ACKNOWLEDGMENTS

I thank Albert Kwok Sing Wong, MA, OTR for his continuous optimism and help.

ABOUT THE AUTHOR

Sharon A. Gutman, PhD, OTR is an assistant professor at Long Island University, where she teaches and conducts research in the Division of Occupational Therapy. She earned her PhD in occupational therapy at New York University with a specialization in neurologic conditions, particularly in traumatic brain injury. Dr. Gutman has been teaching in occupational therapy programs for 8 years. She has been able to draw upon her extensive knowledge of clinical neurologic conditions to write this concise and easy to understand textbook for students in professional health care programs.

SECTION 1

Directional Terminology

DIRECTIONAL TERMINOLOGY

Anterior or Ventral
Refers to the FRONT of the organism. Ventral means the "belly" of a four-legged animal.

Posterior or Dorsal
Refers to the BACK of the organism.

Superior
Refers to the direction ABOVE. One structure is ABOVE another.

Inferior
Refers to the direction BELOW. One structure is BELOW another.

Rostral
Refers to the HEAD of the organism. Also refers to structures that are ABOVE others.

Caudal
Refers to the TAIL of the organism. Also refers to structures that are BELOW others.

Medial
Refers to structures that are close to the MIDLINE of the body.

Lateral
Refers to structures that are further from the MIDLINE of the body.

PLANES OF THE BRAIN

Midsagittal
The MIDSAGITTAL PLANE divides the left and right cerebral hemispheres. This plane divides the brain in half and runs along the MEDIAL LONGITUDINAL FISSURE.

Sagittal
The SAGITTAL PLANES run parallel to the midsagittal plane.

Coronal (also called Frontal or Transverse)
The CORONAL PLANES run perpendicular to the sagittal planes. Coronal planes divide the anterior aspect of the brain from the posterior aspect.

Horizontal
The HORIZONTAL PLANES divide the superior aspect of the brain from the inferior aspect.

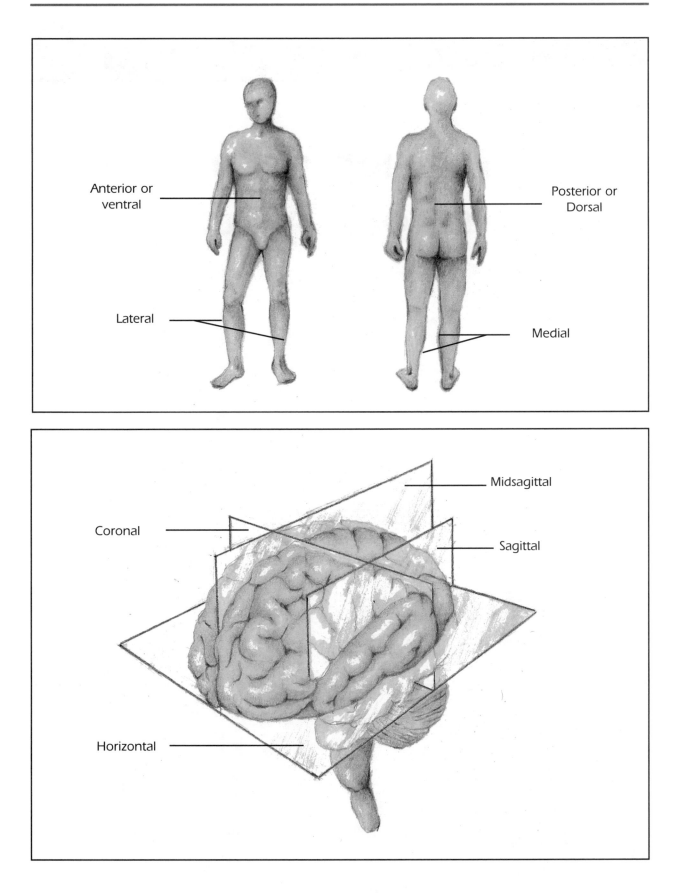

SECTION 2

Division of the Nervous System

1. The NERVOUS SYSTEM is divided into the:
   - CENTRAL NERVOUS SYSTEM (CNS)
   - PERIPHERAL NERVOUS SYSTEM (PNS)

2. The CNS is composed of the:
   - BRAIN and SPINAL CORD

3. Peripheral Nervous System
   - The PNS is composed of the:
     - CRANIAL NERVES (CN)
     - SOMATIC NERVOUS SYSTEM (SNS)
     - AUTONOMIC NERVOUS SYSTEM (ANS)

4. The BRAIN has six major component parts:
   - Cerebral Lobes
   - Cerebellum
   - Basal Ganglia
   - Diencephalon
   - Brainstem
   - Limbic System

5. The ANS is composed of the:
   - PARASYMPATHETIC NERVOUS SYSTEM
   - SYMPATHETIC NERVOUS SYSTEM

6. The SNS is responsible for the Innervation of Skeletal Muscles.

ANS

Innervation of Visceral Muscles and Glands

- ➤ Cardiac Muscle
- ➤ Lungs
- ➤ GI Tract
- ➤ Secretory Glands

Parasympathetic Nervous System	Sympathetic Nervous System
➤ Homeostasis	➤ Arousal
➤ Slowing Body Down	➤ Fight/Flight
➤ Decreased Blood Pressure	➤ Increased Blood Pressure
➤ Decreased Heart Rate	➤ Increased Heart Rate
➤ Peristalsis	➤ Cessation of Peristalsis

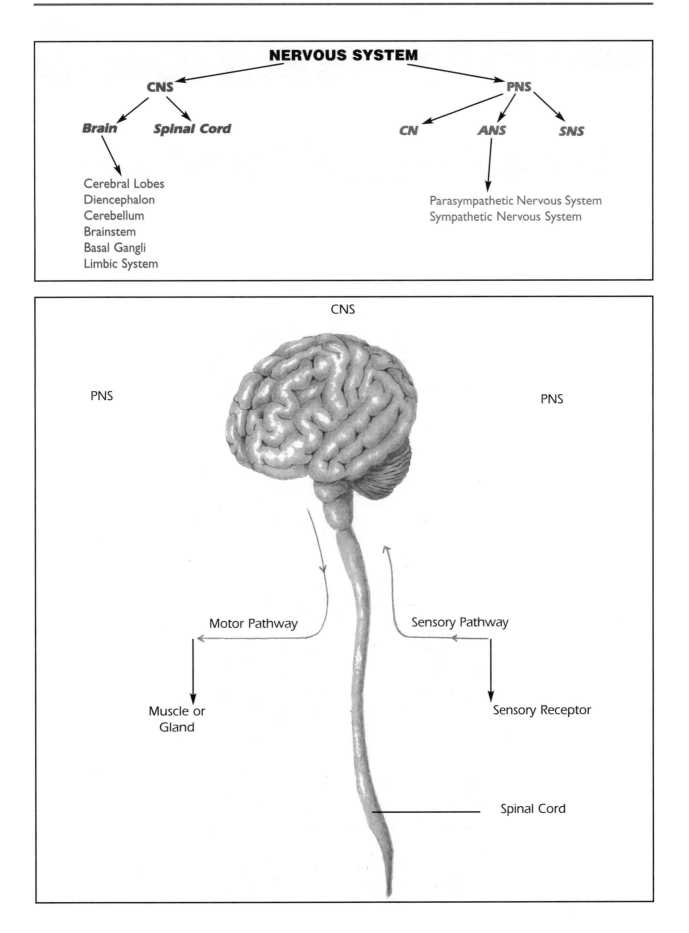

NERVOUS SYSTEM

CNS

Brain   *Spinal Cord*

Cerebral Lobes
Diencephalon
Cerebellum
Brainstem
Basal Gangli
Limbic System

PNS

CN   ANS   SNS

Parasympathetic Nervous System
Sympathetic Nervous System

CNS

PNS   PNS

Motor Pathway   Sensory Pathway

Muscle or
Gland   Sensory Receptor

Spinal Cord

Gross Cerebral Structures

GROSS CEREBRAL STRUCTURES

Gyri (s., Gyrus)
- Gyri are the wrinkles or folds on the surface of the cerebral hemispheres.

Sulci (s., Sulcus)
- Sulci are the valleys or crevices between the Gyri.

Convolutions
- Convolutions are the collective name for the Gyri and Sulci. They are the raised and depressed surfaces of the brain.
- Because brain growth is confined by the skull, the brain folds in on itself as it grows.
- Theorists surmise that the more Gyri and Sulci one has, the larger one's brain surface is, and the more brain capacity one has for brain functions.
- Human brain convolutions are different—like a fingerprint. However, there are certain Sulci and Gyri that are common in all human brains.

Fissure
- A Fissure is a deep groove in the surface of the brain.
- A Fissure is deeper than a Sulcus; one can stick one's fingers into a Fissure. A Sulcus is shallow.

Medial Longitudinal Fissure
- The Medial Longitudinal Fissure separates the right and left cerebral hemispheres.
- This Fissure runs along the Midsagittal Plane.

Central Sulcus (also called the Sulcus of Rolando)
- The Central Sulcus separates the Frontal and Parietal Lobes.
- Also separates the Primary Motor Cortex from the Primary Somatosensory Cortex.

Precentral Gyrus
- The Precentral Gyrus is the Primary Motor Cortex. This handles voluntary motor movement.
- Located just anterior to the Central Sulcus.

Postcentral Gyrus
- The Postcentral Gyrus is the Primary Somatosensory Cortex. It is located just posterior to the Central Sulcus.
- This is the part of our brain that mediates the detection of physical sensation.

Lateral Fissure (also called Fissure of Sylvius)
- This Fissure separates the Temporal Lobe from the Frontal Lobe.

CEREBRAL LOBES

Each hemisphere has four separate lobes.

Frontal Lobes

- Borders of the Frontal Lobes are the Lateral Fissure and the Central Sulcus.
- The Frontal Lobes mediate Cognition (intelligence, problem-solving, short-term memory), Expressive Language, Motor Planning, Mathematical Calculations, and Working Memory.
- The Prefrontal Lobe mediates Executive Functions (organization, planning, sequencing, and motivation), Self-Insight, and Regulation of Emotions.
- The Frontal Lobes develop most after birth. Development is not thought to be complete until late childhood.

Parietal Lobes

- Sit just posterior to the Frontal Lobes.
- The Central Sulcus divides the Parietal Lobes from the Frontal Lobes.
- The Posterior Border is the Parieto-Occipital Sulcus and can be seen on a midsagittal cross-section.
- The Inferior Border is the Temporal Lobe and the Lateral Fissure.
- Function: Sensory Detection, Perception, and Interpretation.

Temporal Lobes

- Most Inferior or Caudal Lobe.
- Poorly defined Posterior Border—Anterior Occipital Lobe.
- Function: Audition or Hearing. Comprehension of Language and Long-Term Memory.
- The INSULA is a Temporal Lobe Structure located deep within the Lateral Fissure. Role in Auditory Processing.

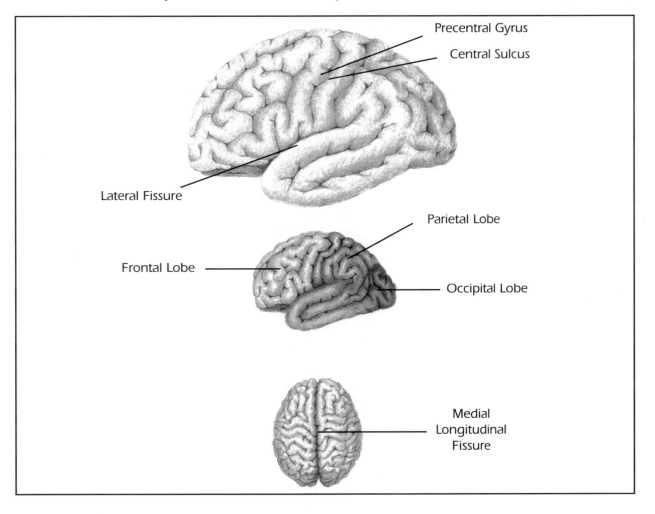

Occipital Lobes
- Most Posterior Lobes.
- Responsible for the Interpretation of Visual Stimuli from the Optic Pathways.

RIGHT VS. LEFT HEMISPHERES

Right Hemisphere
- Largely responsible for the interpretation of PERCEPTUAL and SPATIAL INFORMATION (eg, reading maps, creating music or art).
- Responsible for the interpretation of information that is ABSTRACT and CREATIVE (as opposed to concrete and logical).
- Responsible for the interpretation of TONAL INFLECTIONS in language (as opposed to the concrete meaning of words).
- The Right Hemisphere is responsible for taking the literal interpretation of a story and forming ABSTRACT SYMBOLISM and METAPHORS.
- Also responsible for the interpretation of the EMOTIONAL MESSAGES underlying the concrete meaning of words.
- Controls MOVEMENT on the LEFT side of the body.
- Receives SENSORY information from the LEFT side of the body.

Left Hemisphere
- In people who are right-hand dominant, the Left Hemisphere is usually dominant.
- Plays a large role in Human Language (the expression and interpretation of written and spoken words).
- People with APHASIA often have sustained Left Hemisphere damage.
- Controls MOVEMENT on the RIGHT side of the body.
- Receives SENSORY information from the RIGHT side of the body.

GRAY MATTER VS. WHITE MATTER

- The Cerebral Hemispheres consist of Gray and White Matter.

Gray Matter
- Areas where Gray Matter cover part of the central nervous system (CNS) are called the CORTEX (pl., Cortices).
- Humans have a Cerebral Cortex and a Cerebellar Cortex.
- Gray Matter sits on the surface of the Cerebrum and the Cerebellum.
- It has a grayish or beige appearance because it consists of Nerve Cell Bodies (Nuclei).
- The Gray Matter is Non-Myelinated brain matter. Myelin is a lipid that insulates a nerve and increases conduction velocity.

Ganglia
- Ganglia are a collection of Neural Cell Bodies (or Nuclei) usually located outside of the CNS, or in the peripheral nervous system (PNS).
- Example: Dorsal Root Ganglia.
- The Dorsal Root Ganglia contain the cell bodies of the Sensory Spinal Nerves.

White Matter
- White Matter is located beneath the Gray Matter, in the internal regions of the Cerebrum and Cerebellum.
- White Matter consists of Myelinated Fiber Tracts or Neuronal AXONS.

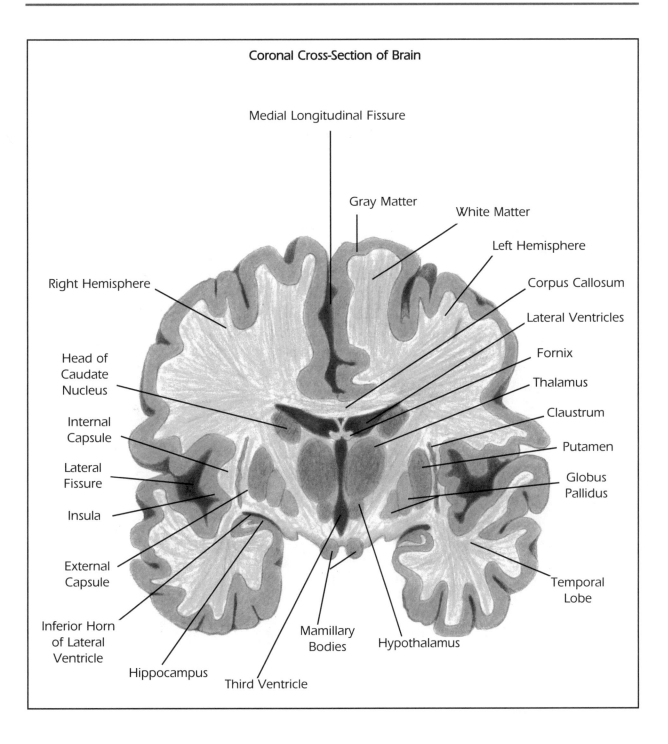

Coronal Cross-Section of Brain

Commissure

- A Commissure is any collection of AXONS (White Matter) that connect one side of the nervous system to the other.
- Example: Corpus Callosum and the Pyramidal Decussation.

DIENCEPHALON

- The Diencephalon is considered to be an old part of the brain in phylogenetic terms.
- The Cortex is considered to be the newest brain region phylogenetically.
- The Diencephalon consists of four structures:
  - Thalamus
  - Hypothalamus
  - Epithalamus
  - Subthalamus

Thalamus

- Means egg-shaped. There are two Thalamic Lobes—one in each Hemisphere.
- The Thalamus contains 26 pairs of nuclei. These primarily receive sensory data from the sensory systems.
- These nuclei then relay the SENSORY data to specific parts of the Cerebral Hemispheres.
- All Sensory Information from the external world is first organized through the Thalamus before it travels to the Cortex for Interpretation.
- The Thalamus is the gateway to the Cortex.
- The Thalamus also receives MOTOR information from the Cerebral Hemispheres and relays it to the Motor Receptors.
- The Thalamus has a role in sleep-wake cycles and consciousness.
- The Thalamus also works with the Reticular Formation to alert the brain to important incoming sensory information and to calm the body down. The Thalamus may work as a screening device for the Cortex.
- Important Thalamic Nuclei include:
  - LATERAL GENICULATE NUCLEUS—Responsible for Visual Processing
  - MEDIAL GENICULATE NUCLEUS—Responsible for Auditory Processing
  - VENTROLATERAL NUCLEUS—Responsible for the organization of Motor Responses
  - VENTRAL POSTEROLATERAL NUCLEUS—Responsible for Tactile-Sensory Processing

Hypothalamus

- Located just anterior and inferior to the Thalamus.
- There are two Hypothalamic Lobes—one in each Hemisphere.
- Function:
- Regulates the Autonomic Nervous System (ANS)
- Releases Hormones from the Pituitary Gland
- Regulates Temperature
- Regulates Hunger
- Regulates Sleep-W   ake Cycles (Circadian Rhythms)
- Works collaboratively with the Limbic System in the Expression of Emotions

Epithalamus

- Located just posterior to the Thalamus and just anterior to the Pineal Gland.
- Very small structure.
- A principal structure of the Epithalamus is the HABENULA—a nucleus at the posterior of the Epithalamus.

Subthalamus

- The Subthalamus is a deep structure.

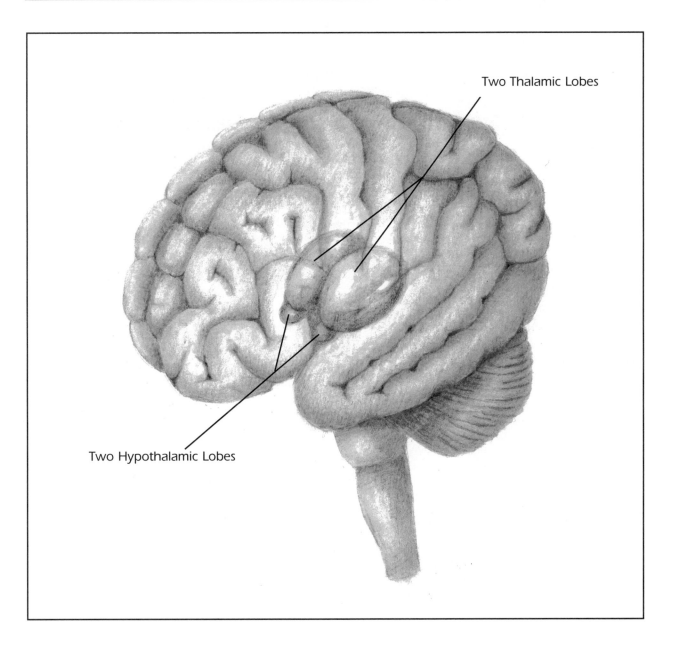

Two Thalamic Lobes

Two Hypothalamic Lobes

- Located just caudal to the Thalamus.
- It is larger than the Epithalamus, and it can be identified in certain coronal cross-sections.

The Epithalamus and Subthalamus have no distinct borders.

OTHER STRUCTURES OF THE DIENCEPHALON

Pituitary Gland
- There is only one Pituitary Gland in the body.
- The Pituitary Gland is an endocrine gland that secretes hormones that Regulate Growth, Reproductive Activities, and Metabolic Processes.
- The Pituitary works collaboratively with the Hypothalamus.

Infundibulum
- The Infundibulum is the Stalk that extends from the Hypothalamus and holds the Pituitary Gland.

Pineal Gland
- The Pineal Gland is a midline structure, located just posterior to the Thalamus.
- There is only one Pineal Gland in the body.
- The Pineal Gland is innervated by the ANS.
- Has a role in sexual and hormonal functions and in sleep-wake cycles.

Posterior Commissure
- Connects the right and left halves of the Diencephalon.
- Located just above the Superior Colliculi.
- Allows communication between the Hemispheres if the Corpus Callosum is lesioned or removed (secondary to pathology).

Anterior Commissure
- Connects the Olfactory Bulb to the Amygdala.
- Located in the anterior Thalamus and passes through the head of the Caudate Nucleus.
- Allows communication between the Hemispheres if the Corpus Callosum is lesioned.

Interthalamic Adhesion
- Allows communication between the two Thalamic Lobes.
- Located centrally in the Thalamus.

Septum Pellucidum
- A sheath-like cover that extends over the medial wall of each Lateral Ventricle.
- May have a role in the processing of Emotion along with the Limbic System.

STRUCTURES LOCATED NEAR THE DIENCEPHALON (BUT ARE NOT PART OF THE DIENCEPHALON)

Corpus Callosum
- The Corpus Callosum is the largest Commissure in the brain.
- It allows the Right and Left Cerebral Hemispheres to communicate with each other.
- The Corpus Callosum arches around the Anterior Horn of the Lateral Ventricles.

Optic Chiasm
- A Chiasm is a crossing-over point.
- The Optic Chiasm is a cross-shaped connection between the optic nerves.
- It is a midline structure, located at the base of the brain, just superior to the Pituitary Gland.

Internal Capsule

- The Internal Capsule is a Large Fiber Bundle that connects the Cerebral Cortex with the Diencephalon.
- All Descending Motor Messages from the Cortex travel through the Internal Capsule to the Thalamus, Brainstem, Spinal Cord, and to the Skeletal Muscles.
- Sensory information from the sensory receptors ascends in the Spinal Cord, through the Brainstem, to the

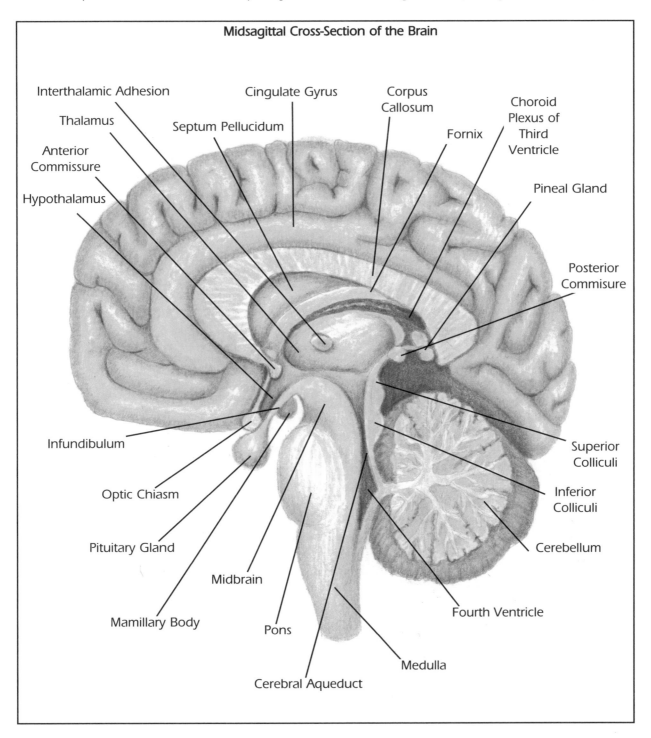

Midsagittal Cross-Section of the Brain

Interthalamic Adhesion
Cingulate Gyrus
Corpus Callosum
Choroid Plexus of Third Ventricle
Thalamus
Septum Pellucidum
Fornix
Anterior Commissure
Hypothalamus
Pineal Gland
Posterior Commisure
Infundibulum
Superior Colliculi
Optic Chiasm
Inferior Colliculi
Pituitary Gland
Cerebellum
Midbrain
Mamillary Body
Pons
Fourth Ventricle
Cerebral Aqueduct
Medulla

Thalamus, through the Internal Capsule, and to the Primary Somatosensory Cortex.

Mammillary Bodies

- Two protrusions that sit within the Interpeduncular Fossa on the Anterior Side of the Midbrain.
- The Mammillary Bodies are Nuclei Groups that form attachments with the Hypothalamus and Fornix and may play a role in the processing of emotion.

BRAINSTEM

- The Brainstem is composed of three basic structures:
  - Midbrain
  - Pons
  - Medulla
- The Brainstem controls Vegetative Functions including:
  - Respiration
  - Cough and Gag Reflex
  - Pupillary Response
  - Swallowing Reflex

Midbrain

- Most Rostral structure of the Brainstem.
- Sits atop of the Pons.
- Just Inferior to the Thalamus.
- The Midbrain has a role in Automatic Reflexive Behaviors dealing with Vision and Audition.

External Structures of the Midbrain

Cerebral Peduncles

- Large Fiber Bundles located on the Anterior Surface of the Midbrain.
- Carries descending Motor Tracts from the Cerebrum to the Brainstem.

Interpeduncular Fossa

- This is the indentation between the pair of Cerebral Peduncles.
- The Mammillary Bodies sit within the Interpeduncular Fossa on the ANTERIOR Aspect of the Midbrain.

Superior and Inferior Colliculi

- The Superior and Inferior Colliculi sit on the POSTERIOR surface of the Midbrain—just beneath the Posterior Commissure.
- The Superior Colliculi are located just above the Inferior Colliculi.
- The Superior Colliculi are a pair of Relay Centers for Vision. They communicate directly with the Thalamic Nuclei (Lateral Geniculate Nuclei) that process Visual Stimuli.
- The Inferior Colliculi are a pair of Relay Centers for Audition. They communicate directly with the Thalamic Nuclei (Medial Geniculate Nuclei) that process Auditory Stimuli.

Internal Structures of the Midbrain (Seen on Cross-Section)

Cerebral Aqueduct (also called Aqueduct of Sylvius)

- The Cerebral Aqueduct is part of the Ventricular System. It connects the 3rd and 4th Ventricles. Cerebrospinal Fluid (CSF) flows through the Cerebral Aqueduct.

Superior and Inferior Colliculi

- Can be seen in cross-section.

Cerebral Peduncles

- Has an Inner and Outer Coat.

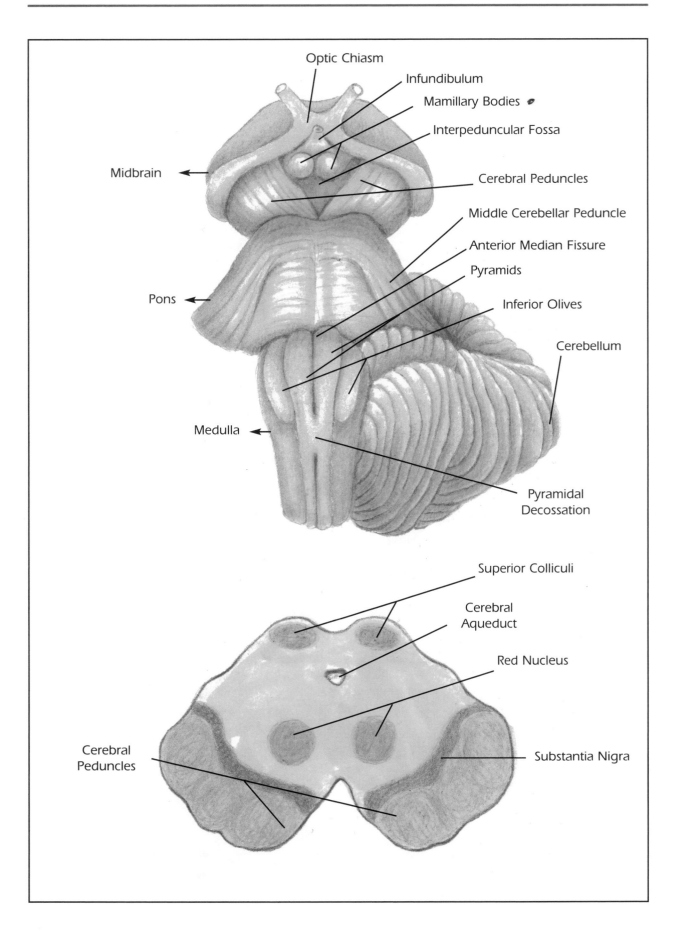

Optic Chiasm

Infundibulum

Mamillary Bodies

Interpeduncular Fossa

Midbrain

Cerebral Peduncles

Middle Cerebellar Peduncle

Anterior Median Fissure

Pyramids

Pons

Inferior Olives

Cerebellum

Medulla

Pyramidal
Decossation

Superior Colliculi

Cerebral
Aqueduct

Red Nucleus

Cerebral
Peduncles

Substantia Nigra

- Outer Coat—consists of the CRUS CEREBRI.
- Inner Coat—consists of the RED NUCLEUS and the SUBSTANTIA NIGRA.

Tegmentum
- The Substantia Nigra and the Red Nucleus are collectively called the Tegmentum.

Tectum
- The Superior and Inferior Colliculi are collectively called the Tectum.

PONS

- Located just Caudal to the Midbrain and Rostral to the Medulla.
- The Pons is a relay system between the Spinal Cord, Cerebellum, and the Cerebrum.
- It largely mediates motor information on an unconscious level—shifting weight to maintain balance and making fine motor adjustments in one's muscles to perform precise coordinated limb movement.

External Structures of the Pons

Cerebellar Peduncles
- Largely carries Sensory Information from the Pons to the Cerebellum about the body's position in space.
- Each Cerebellar Peduncle is paired—one on each side.
- There are three Cerebellar Peduncles:
  - MIDDLE CEREBELLAR PEDUNCLE:
    - Carries Sensory Information about the body's position in space to the Cerebellum
  - INFERIOR CEREBELLAR PEDUNCLE:
    - Located in the Pons and Medulla
    - Carries Sensory Information about the body's position in space from the Pons/Medulla to the Cerebellum
    - The Cerebellum then analyzes this sensory information and makes decisions about how to readjust the body for precision movement and balance
    - The Cerebellum then sends its decision to the Thalamus via the Superior Cerebellar Peduncles. The Thalamus sends this motor information back down through the Brainstem to the Spinal Cord, and to the Skeletal Muscles
  - SUPERIOR CEREBELLAR PEDUNCLE
    - Located in the Pons
    - Carries Sensory Information from the Pons to the Cerebellum
    - Also carries Sensorimotor Information from the Cerebellum to the Thalamus

Internal Structures of the Pons

4th Ventricle
- Located in the Posterior Pons.
- The 4th Ventricle is a continuation of the Cerebral Aqueduct.
- The 4th Ventricle leads into the Central Canal of the Spinal Cord.

Corticospinal Tracts
- These are Descending Motor Tracts from the Primary Motor Cortex.

Middle and Superior Cerebellar Peduncles
- Can be seen in a cross-section of the Pons.
- Carries Sensory Information from the Pons to the Cerebellum.

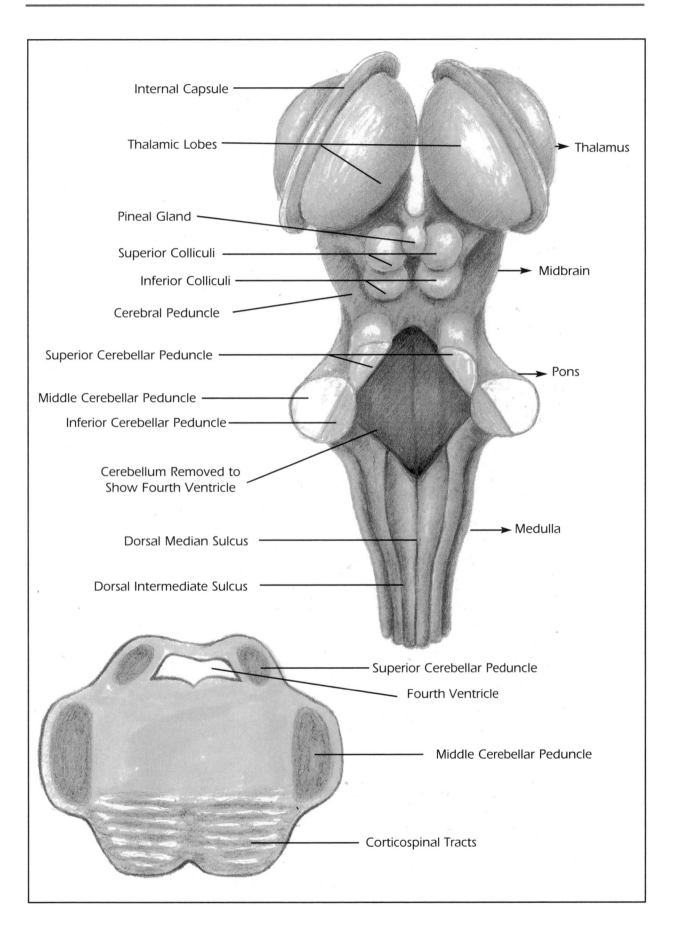

Internal Capsule

Thalamic Lobes

Thalamus

Pineal Gland

Superior Colliculi

Inferior Colliculi

Midbrain

Cerebral Peduncle

Superior Cerebellar Peduncle

Pons

Middle Cerebellar Peduncle

Inferior Cerebellar Peduncle

Cerebellum Removed to
Show Fourth Ventricle

Medulla

Dorsal Median Sulcus

Dorsal Intermediate Sulcus

Superior Cerebellar Peduncle

Fourth Ventricle

Middle Cerebellar Peduncle

Corticospinal Tracts

MEDULLA

- The Medulla is just caudal to the Pons.
- Because the Medulla is long it is divided into the ROSTRAL and CAUDAL Medulla.
- The Medulla carries descending Motor Messages from the Cerebrum to the Spinal Cord.
- Also carries ascending Sensory Messages from the Spinal Cord to the Cerebrum.

External Structures of the Medulla—Anterior Side

Anterior Median Fissure

- The Anterior Median Fissure divides the Medulla into equal right and left halves.
- This Fissure continues all the way down the Spinal Cord.

Pyramids

- The Pyramids are two large structures that are divided by the Anterior Median Fissure.
- The Pyramids are Motor Tracts or Fiber Bundles that carry Descending Motor Information from the Cortex to the Spinal Cord.

Pyramidal Decussation

- This is the Crossing-Over Point where Motor Fibers from the left Cortex cross to the right side of the Spinal Cord. Motor Fibers from the right side of the Cortex cross to the left side of the Spinal Cord.
- This is why the Right Cerebral Hemisphere controls the Left side of the Body and the Left Cerebral Hemisphere controls the Right side of the Body.

Inferior Olives

- The Olives are located just lateral to each Pyramid.
- The Olives are relay nuclei that carry ascending Sensory Information to the Cerebellum. The sensory data pertain to the body's position in space.

External Structures of the Medulla—Posterior Side

Dorsal Median Sulcus

- The Dorsal Median Sulcus is the Midline that divides the posterior Medulla into equal left and right sides.
- It is not as well-defined as the Anterior Median Fissure.

Dorsal Intermediate Sulcus

- These Sulci are located just lateral to the Dorsal Median Sulcus.

Internal Structures of the Medulla (Seen in Cross-Section)

Rostral Medulla

- 4th Ventricle
- Pyramids
- Inferior Olivary Nuclei
- Inferior Cerebellar Peduncles

Caudal Medulla

- Central Canal (this is where the end of the 4th Ventricle meets the Spinal Cord)
- Pyramids
- Fasciculus Gracilis and Cuneatus (these are ascending Sensory Tracts)
- Nucleus Gracilis and Cuneatus (these are where the nuclei or cell bodies of the Fasciculus Gracilis and Cuneatus are located)

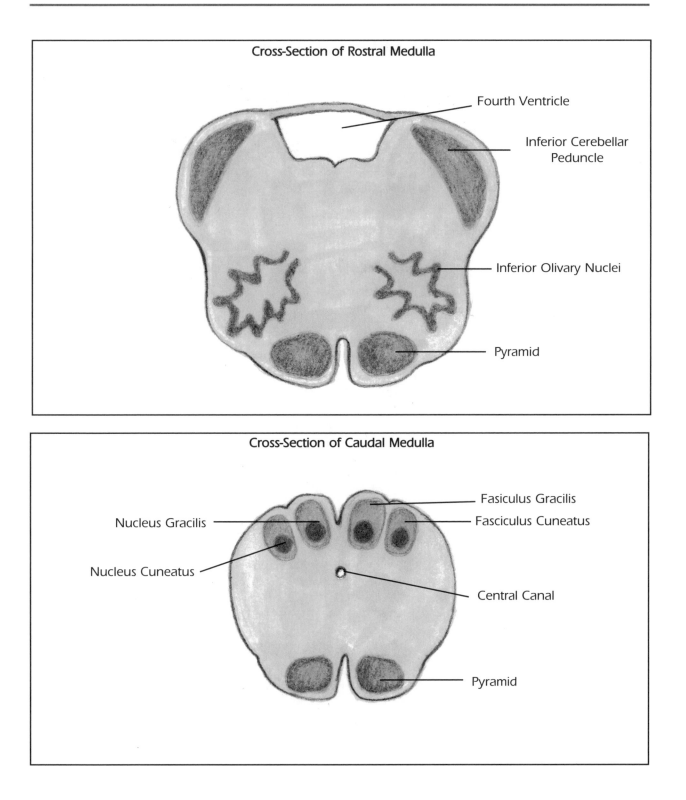

Cross-Section of Rostral Medulla

Fourth Ventricle

Inferior Cerebellar Peduncle

Inferior Olivary Nuclei

Pyramid

Cross-Section of Caudal Medulla

Fasiculus Gracilis

Fasciculus Cuneatus

Nucleus Gracilis

Nucleus Cuneatus

Central Canal

Pyramid

RETICULAR FORMATION

- Diffusely located in the Brainstem.
- Consists of two systems:
  - Reticular Activating System
  - Reticular Inhibiting System

Reticular Activating System

- Brainstem center that is involved in states of Wakefulness.
- Plays a role in alerting the Cortex to attend to important Sensory Stimuli.

Reticular Inhibiting System

- Brainstem center that is involved in states of Unconsciousness such as Sleep, Stupor, or Coma.

BASAL GANGLIA

- The Basal Ganglia is an Unconscious Motor System that operates on a SUBCORTICAL Level.
- Specifically mediates Stereotypic or Automatic Motor Patterns such as those involved in walking, riding a bike, and writing.
- These are activities that are initially learned by using the Cortex to think about how to perform such motor movements.
- Once learned, such motor patterns are stored subcortically. They become integrated by the Basal Ganglia.
- The Basal Ganglia consists of three structures:
  - CAUDATE NUCLEUS
  - PUTAMEN
  - GLOBUS PALLIDUS
- These are clusters of neurons within the Cerebrum.
- The Basal Ganglia are the only Ganglia in the CNS.

Caudate Nucleus

- The Caudate is an arch-shaped structure that follows the arc of the Fornix and Lateral Ventricles.
- The Caudate has a Head, Body, and Tail. The Tail attaches to the Amygdala of the Limbic System.
- Involved in the PLANNING and EXECUTION of a particular Automatic Movement Pattern.
- Also involved in the Evaluation of that Movement's Appropriateness.
- The Caudate has strong connections with the Frontal Lobe and interacts with it in MOTOR PLANNING.
- The Caudate also plays a role in the INHIBITORY CONTROL of MOVEMENT. The Caudate acts like a brake on certain motor activities.
- When the Brake is not working—when the Caudate is damaged—Extraneous, Purposeless Movements appear.
- Example: Tics, Tardive Dyskinesias (tongue protrusions, facial grimacing, lip smacking).
- Tardive means that the disorder occurred after chronic use of certain drugs that affect the Caudate Nucleus.

Putamen and Globus Pallidus

- The Putamen and Globus Pallidus are EXCITATORY Structures.
- They are located just lateral to the Internal Capsule.

Extrapyramidal System

- The Basal Ganglia is considered to be an Extrapyramidal System—or a motor system that does not use the Pyramids to send motor messages to the skeletal muscles.

Corpus Striatum

- Collective name for the CAUDATE, PUTAMEN, and GLOBUS PALLIDUS.

Neostriatum

- Collective name for the CAUDATE and PUTAMEN.

Lenticular Nucleus
- Collective name for the GLOBUS PALLIDUS and PUTAMEN.

Paleostriatum
- Another name for the GLOBUS PALLIDUS—because of its striped appearance.

Claustrum
- The Claustrum is NOT considered to be part of the Basal Ganglia although it is a group of nuclei located just Lateral to the Extreme Capsule and just Medial to the Insula.
- Its function is unknown.

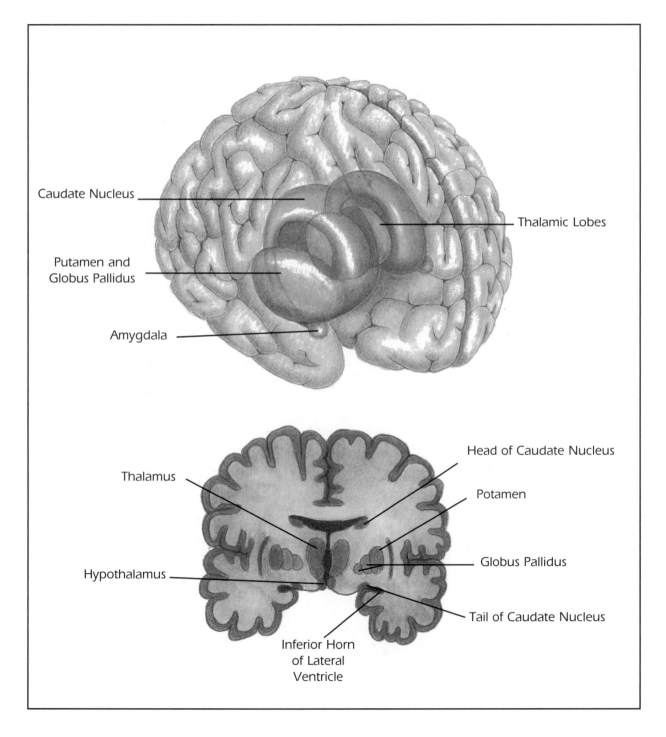

Caudate Nucleus

Thalamic Lobes

Putamen and
Globus Pallidus

Amygdala

Thalamus

Head of Caudate Nucleus

Potamen

Globus Pallidus

Hypothalamus

Tail of Caudate Nucleus

Inferior Horn
of Lateral
Ventricle

CEREBELLUM

- Cerebellum means "Little Brain," and in many ways it is like a brain unto itself.
- The Cerebellum has two Hemispheres that are connected by a Vermis—much like the Corpus Callosum connects the Cerebral Hemispheres.
- It has a Cerebellar Cortex with an outer coat of Gray Matter and an inner core of White Matter.
- The Cerebellum has three Lobes.
- Major Function: PROPRIOCEPTION or the unconscious awareness of the body's position in space.
- The Cerebellum is a Sensory and Motor System. It receives Sensory Information from joint and muscle receptors concerning the body's position. The Cerebellum uses this Sensory Information to make decisions about how to adjust the body for the coordinated, precision control of movement and balance.
- These decisions are made on an unconscious level—the messages never reach the Cortex for conscious awareness.

External Structures of the Cerebellum

- Two Hemispheres—each has three Lobes.

Flocculonodular Lobe

- Also called the ARCHICEREBELLUM (ie, ancient brain).
- Considered to be the oldest part of the Cerebellum in phylogenetic terms.
- Believed to have developed in organisms without limbs.
- Role in Trunk Control, Postural Reflexes, and Balance.

Anterior Lobe

- Also called the PALEOCEREBELLUM (ie, old brain).
- Evolved in organisms with limbs.
- Functions in Extremity Control, Postural Adjustments, and Stereotypic Movement Patterns.

Posterior Lobe

- Also called the NEOCEREBELLUM (ie, new brain).
- Evolved in organisms with a Cortex.
- Role in Motor Planning (Praxis), and the Precise Timing and Coordination of Multiple Muscle Groups.

Three Primary Fissures

- Primary Fissure—separates the Anterior and Posterior Lobes.
- Posterolateral Fissure—divides the Flocculonodular and Posterior Lobes.
- Horizontal Fissure—divides the Posterior Lobe in Half.

Cerebellar Peduncles

- The Peduncles are composed of Axons or fibers that travel between the Cerebellum and the Brainstem (Superior and Middle Cerebellar Peduncles travel through the Pons; Inferior Cerebellar Peduncle travels through the Medulla).
- Located in the ANTERIOR Lobe of the Cerebellum.

Vermis

- The Vermis is a midline structure that has a role in the integration of information used by the right and left Cerebellar Hemispheres.

Internal Structures of the Cerebellum

Cortex with Folia

- The outer covering of the Cerebellum is lined with Gray Matter that has many folds—called Folia.

White and Gray Matter

- The Interior of the Cerebellum consists of a central gray mass of White Matter surrounded by Gray Matter.
- There are also some smaller masses of Gray Matter within the Central White Core. These are the Cerebellar Nuclei.

Four Pairs of Nuclei
- There are four pairs of Nuclei—one in each Hemisphere.
  - Dendate Nuclei (looks similar to the Inferior Olivary Nuclei of the Medulla)
  - Globose Nuclei
  - Emboliform Nuclei
  - Fastigial Nuclei

Vermis
- The Vermis can also be seen in the interior of the Cerebellum.

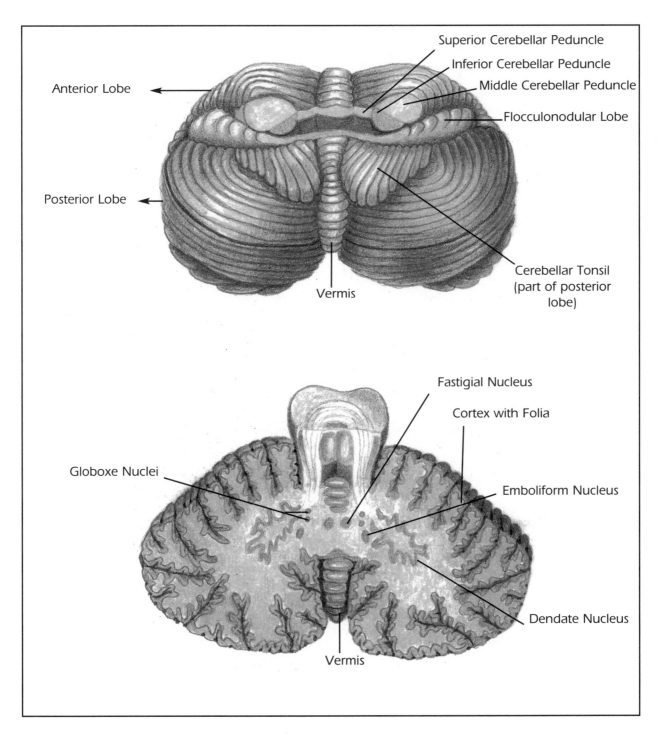

LIMBIC SYSTEM

- Phylogenetically, the Limbic System is considered to be a very old part of the Brain.
- Located deep within the core of the Brain.
- The Limbic System appears to be the source of our Raw Emotions before they are modulated by the Frontal Lobes.
- The Limbic System is also a Storehouse for Long-Term Memories—particularly memories that have a strong emotional component.

Cingulate Gyrus

- Most medial and deepest Gyrus in the Frontal and Parietal Lobes.
- Sits right above the Corpus Callosum.
- Has vast connections to the other structures within the Brain's emotional Limbic System.
- Role in decision-making regarding which actions to take in response to sensory data.

Parahippocampal Gyrus

- Most medial and deepest Gyrus in the Temporal Lobes.
- Folds back on itself at its anterior end to become the UNCUS.
- Relays information between the Hippocampus and other Cerebral areas—particularly the Frontal Lobes.
- The Parahippocampal Gyrus functions when we compare a present event to an event stored in Long-Term Memory in order to decide how to handle a present situation.

Uncus

- The Uncus is the Bulb-like, anterior end of the Parahippocampal Gyrus.

Fornix

- The Fornix Bodies are a pair of arched-shaped fibers that begin in the Uncus and wrap around to the Mamillary Bodies.
- The Fornix is a relay system for messages generated by the Limbic System.

Amygdala

- An almond-shaped nucleus in the anterior Temporal Lobe.
- Attaches to the Caudate Nucleus.
- Role in the mediation of Fear and Anger.
- Also may play a role in the perception of Social Cues and the generation of Feelings of Empathy.
- The Amygdala has been found to be significantly smaller in some people with autism who have difficulty perceiving and interpreting social cues.

Olfactory Bulb and Tract

- Also known as Cranial Nerve 1.
- The Olfactory Tract connects to Limbic System Structures—it travels directly to the Hippocampus.
- This connection accounts for the deep association between specific odors and Long-Term Memories that hold emotional significance.

Hippocampus

- The Hippocampus is located within the Parahippocampal Gyrus.
- In a Coronal Section, the Hippocampus looks like a seahorse (Hippocampus means "seahorse").
- One of the Major Storehouses in the Brain for Long-Term Memory.

Limbic System

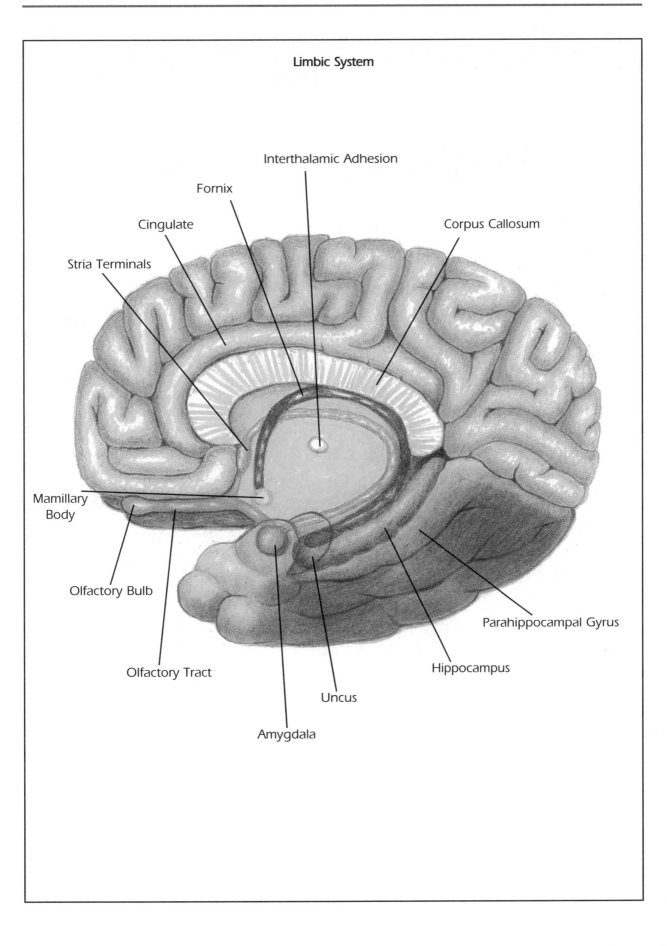

Interthalamic Adhesion

Fornix

Cingulate

Corpus Callosum

Stria Terminals

Mamillary
Body

Olfactory Bulb

Parahippocampal Gyrus

Olfactory Tract

Hippocampus

Uncus

Amygdala

Ventricular System

VENTRICULAR SYSTEM

- The Brain begins as a Flat Plate that fuses into a Tube.
- The Space within the Tube becomes the Ventricular System.
- The Tube itself is the Brain; the Ventricles are the Hollow Spaces in the Brain that Contain Cerebrospinal Fluid (CSF).

There Are Four Ventricles in the Brain

- One Pair of Lateral Ventricles—One in each Hemisphere.
- One 3rd Ventricle.
- One 4th Ventricle.

Two Lateral Ventricles

- There is one Lateral Ventricle in each Hemisphere.
- The Lateral Ventricles are Divided by the Septum Pellucidum—a thin partition covering the medial wall of each Lateral Ventricle.
- Each Lateral Ventricle has three Horns:
  - Anterior Horn—projects into the Frontal Lobe
  - Inferior Horn—projects into the Temporal Lobe
  - Posterior Horn—projects into the Occipital Lobe

One 3rd Ventricle

- The 3rd Ventricle is surrounded by the Diencephalon.
- The Thalamic Lobes form the Walls of the 3rd Ventricle.
- The Hypothalamic Lobes form the Floor of the 3rd Ventricle.

One 4th Ventricle

- The 4th Ventricle is located between the Pons, Rostral Medulla, and the Cerebellum.

Choroid Plexes

- The Choroid Plexes are the Vascular Structures in the Brain that protrude into the Ventricles and Produce the CSF.
- All of the Ventricles contain Choroid Plexes, but the Lateral Ventricles contain the most.

Cerebral Aqueduct (also called Aqueduct of Sylvius)

- The Cerebral Aqueduct is a narrow channel that descends through the Midbrain.
- It connects the 3rd and 4th Ventricles.
- Common Site of Blockage.

Central Canal

- Begins in the Caudal Medulla and descends all the way down the Spinal Cord.
- Contains CSF.
- The Central Canal connects the Ventricular System with the Spinal Cord.

Foramina of Monro
- There are two Foramina of Monro—one in each Hemisphere.
- They are small channels that connect the Lateral Ventricles with the 3rd Ventricle.

Foramina of Magendie (also called Median Aperture)
- There is only one Foramina of Magendie. This is an opening in the 4th Ventricle (in the Rostral Medulla).
- The Foramina of Magendie opens to the SUBARACHNOID SPACE below the Cerebellum. This space is located above the Brain and beneath the Skull.
- The SUBARACHNOID SPACE is the space between the ARACHNOID MEMBRANE and the PIA MATTER.
- This is also a Potential Site of CSF Blockage.

Foramina of Luschka (also called Lateral Aperture)
- There are two Foramina of Luschka. These are Openings in the 4th Ventricle (in the Pons).
- Opens to the Subarachnoid Space.
- Potential Sites of Blockage.

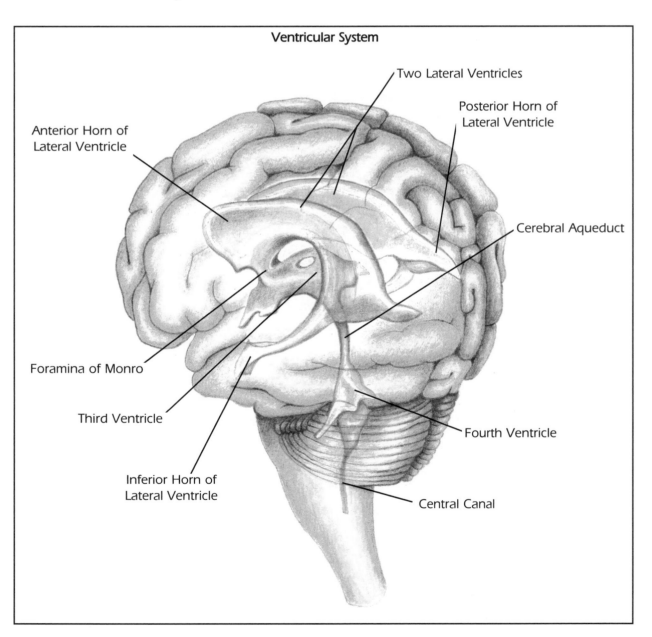

Ventricular System

Two Lateral Ventricles

Posterior Horn of Lateral Ventricle

Anterior Horn of Lateral Ventricle

Cerebral Aqueduct

Foramina of Monro

Third Ventricle

Fourth Ventricle

Inferior Horn of Lateral Ventricle

Central Canal

CEREBROSPINAL FLUID

- The CSF is a clear, colorless fluid that bathes and nourishes the Brain and Spinal Cord.

Arachnoid Villae

- The CSF is Reabsorbed in the Arachnoid Villae.
- The Arachnoid Villae are projections of the Arachnoid Matter into the Dura Matter.

Cerebrospinal Fluid Pressure

- CSF Maintains a Constant Circulatory Pressure—unless a problem occurs.
- The Formation of CSF is Independent of the Pressure.
- This is Important with Regard to HYDROCEPHALUS.
- Even if the CSF Pressure increases, the CSF continues to be produced.
- There is NO Neurologic Mechanism that Detects too much CSF.

Composition of Cerebrospinal Fluid

- The Composition of CSF is used for Diagnostic Purposes to Identify Disease Processes.
- Physicians examine the Rate of Pressure and the Composition.
- Example: SPINAL TAP is a in which the Spinal Cavity is Punctured with a Needle to Extract the CSF for Diagnostic Purposes.

Flow of Cerebrospinal Fluid

- The CSF is Produced in the CHOROID PLEXES, travels through the Ventricles, Subarachnoid Space, and the Spinal Cord, and is then reabsorbed by the Arachnoid Villae.

Function of the Cerebrospinal Fluid

- PROTECTION OF THE BRAIN—the Fluid acts like a Shock Absorber.
- EXCHANGE OF NUTRIENTS AND WASTE—CSF plays a role in the transfer of substances between the Blood and the Nervous Tissue.
- DIAGNOSIS—the CSF is examined for its Rate of Pressure and its Fluid Composition.
- TRANSPORT OF HORMONES—the CSF has a role in the transport of some Hormones throughout the CNS.

HYDROCEPHALUS

- Hydrocephalus is a Buildup of Pressure and Fluid that results in COMPRESSION OF NEURAL TISSUE and ENLARGEMENT OF THE VENTRICLES.
- It can occur in both Infants and Adults.

Congenital Hydrocephalus in Infants

Etiology

- Blockage (particularly in the Foramen).
- Excessive Production of CSF for Unknown Reasons.
- Meningitis: Causing Adhesions and resultant Blockages in the Subarachnoid Space.
- Tumors of the Choroid Plexes: Causing Excessive CSF Production.

Common Sites of Blockage

- Cerebral Aqueduct.
- Foramen of Lushka.
- Foramen of Magendie.

Pathological Effects

- Infant's Skull Expands to Accommodate the increased fluid.
- Compression of Neural Tissue.
- Enlarged head (because the skull is still malleable due to the Fontanels).

Treatment

- If the Hydrocephalus was caused by a blockage, then treatment requires a SHUNT, or tube, that bypasses the blockage.

- If there is excessive production of CSF, treatment requires a SHUNT usually placed from the 4th Ventricle to the Abdomen.

Adult Onset Hydrocephalus

Etiology

- Tumors.
- Meningitis.
- Hemorrhage.
- Unknown Causes.

Pathological Effects

- Enlarged Ventricles and rapid Compression of Neural Tissue—there is no place for the fluid to go.
- The increased fluid impinges upon Brain Tissue.
- This condition is life-threatening unless treatment occurs quickly.

Treatment

- Treatment for an adult usually involves SHUNT placement (to the ABDOMEN) to drain the excess fluid.

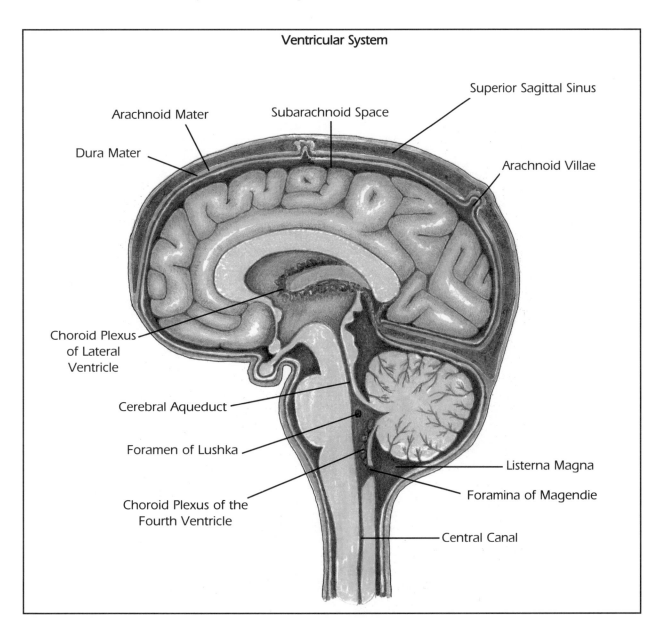

Ventricular System

Arachnoid Mater

Subarachnoid Space

Superior Sagittal Sinus

Dura Mater

Arachnoid Villae

Choroid Plexus of Lateral Ventricle

Cerebral Aqueduct

Foramen of Lushka

Listerna Magna

Foramina of Magendie

Choroid Plexus of the Fourth Ventricle

Central Canal

SECTION 5

The Cranium

SKULL

- The Skull is the Bony Framework of the Head.
- Supports, Anchors, and Protects the Brain.
- Composed of 14 Bones of the Face, 28 Adult Teeth, and 8 Cranial Bones.

Cranium

- The Cranium is the portion of the skull that encloses the Brain.
- Consists of Eight Separate Fused Bones: Frontal Bone, Occipital Bone, Sphenoid Bone, Ethmoid Bone, Two Temporal Bones, and Two Parietal Bones.

Suture Lines

- The Suture Lines are Junctions between the Skull Bones. These are areas where the Bones have fused.
- The Cranial Sutures begin to fuse at 2 months and are complete at 18 months.

Coronal Suture

- Runs along the CORONAL Plane.
- Connects the Frontal Bone with the Parietal Bones.

Sagittal Suture

- Runs along the Midsagittal Plane.
- Connects the Two Parietal Bones.

Lambdoid Suture

- Connects the Two Parietal Bones to the Occipital Bone.

Fontanels

- The Fontanels are unossified spaces or soft spots located between the cranial bones of a fetus and newborn.
- These allow the skull to expand to accommodate the growing brain.
    - ANTERIOR FONTANEL
    - POSTERIOR FONTANEL
    - SPHENOID FONTANEL
    - MASTOID FONTANEL

Floor of the Cranial Cavity

- Where the undersurface of the Brain sits—holds the anterior-inferior aspect of the Frontal Lobes, the inferior aspect of the Temporal Lobes and the inferior aspect of the Cerebellum.

Fossa

- The undersurface of the Brain sit in Three Cranial Sections or FOSSA:
- ANTERIOR CRANIAL FOSSA: Primarily supports the Frontal Lobes.
- MIDDLE CRANIAL FOSSA: Supports the Anterior-Inferior Temporal Lobes and the Diencephalon.
- POSTERIOR CRANIAL FOSSA: Supports the Cerebellum.

Sharp Edges of the Fossa
- The Fossa are problematic in Brain Injuries caused from motor vehicle accidents (MVA).
- Sharp Edges of the Fossa shear Brain Tissue in MVAs.

Foramina
- The Foramina are openings in the Skull for the passage of Blood Vessels and Nerves.

Foramen Magnum
- The Foramen Magnum is the largest foramina in the skull.
- It is the opening through which the Brainstem connects with the Spinal Cord.
- Located in the Occipital Bone.

Function of the Cranial Bones
- Provide PROTECTION for the Brain.

Skull Fractures
- One of the weakest cranial sutures is the Pterion Suture—a site that joins the Frontal, Parietal, Temporal, and Sphenoid Bones.
- The Pterion often fractures upon strong impact causing penetration of bone fragments to enter the Brain; Cerebral Arteries are easily ruptured as a result.
- Frontal and Parietal Bone Fractures are often seen as a result of MVAs.

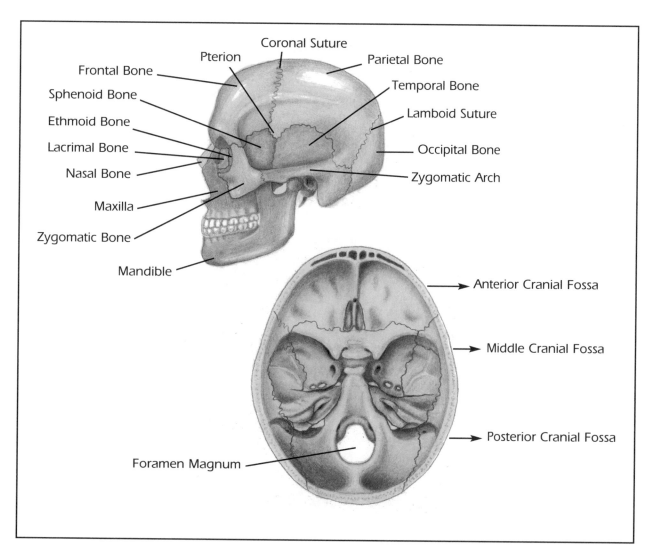

The Meninges

LOCATION

- The Meninges are located between the Skull and Brain, and they cover the Spinal Cord. They form a seal around the Central Nervous System (CNS).
- There are Three Layers of Meninges: Dura Mater, Arachnoid Mater, and Pia Mater.
- These are the layers between the Skull and Brain: SKULL, EPIDURAL SPACE, DURA MATER, SUBDURAL SPACE, ARACHNOID MATER, SUBARACHNOID SPACE, PIA MATER, BRAIN.

DURA MATER

- The Dura is the Outermost Meningeal Layer.
- It is a very tough and thick membrane that is attached to the Inner Surface of the Cranium.
- The Dura has two projections that Extend into the Brain:
  - FALX CEREBRI: Extends into the Medial Longitudinal Fissure.
  - TENTORIUM: The Horizontal Shelf of Dura that sits between the Occipital Lobe and the Cerebellum.
- The FALX CEREBRI and TENTORIUM decrease Linear and Rotary Forces on the Brain.

Dural Sinuses

- The Dural Sinuses are openings for Blood Vessels and Nerves in the Dura.
- Located above the Frontal and Parietal Lobes.
- The Sinuses function as a Circulatory System.
- Cerebral Veins empty into the Sinuses.
- They also receive Cerebrospinal Fluid (CSF) from the Subarachnoid Space.
- These fluids are then returned to their general circulatory systems.

Blood Supply of the Dura

- The Blood Supply of the Dura comes from the MIDDLE MENINGEAL ARTERY.
- This artery often ruptures during Head Injury resulting in Bleeds in the Subdural Space.
- This leads to SUBDURAL HEMATOMA.
  - Subdural Hematomas can also result from Cerebrovascular Accident.
  - Bleeds in the Subdural Space cause Increased Cranial Pressure that Compresses Neural Tissue.
  - This is fatal if not treated quickly.

Dural Neuronal Innervation

- The Dura has Neuronal Innervation—it is innervated by the nervous system and can experience physical sensation.
- The Brain has no way of detecting Pain—it has NO Pain Receptors.
- Some Headaches are often caused by the constriction of Meningeal Membranes.

ARACHNOID MATER

The Arachnoid Mater is the Middle Meningeal Layer.
- Located just below the Subdural Space.
- The Arachnoid looks like a Spider Web (arachnoid means spider).

- Protects the Brain and acts as a seal around the CNS.

Subarachnoid Space

Beneath the Arachnoid is the SUBARACHNOID SPACE and it holds the CSF.

Cisterns

The CISTERNS are openings or large spaces in the Subarachnoid Space.

Cisterna Magna (also called the Cerebellar Medullary Cistern)

- This is the largest Subarachnoid Cistern.
- Located between the Cerebellum and the Medulla.
- Often used as a SHUNT PLACEMENT.

PIA MATER

- The Pia is the Deepest Meningeal Layer.
- Located right on the gyri and sucli of the Brain and on the Spinal Cord.

BLOOD-BRAIN BARRIER

- The Blood-Brain Barrier consists of the Meninges, the Protective Glial Cells, and the Capillary Beds of the Brain.
- Responsible for the Exchange of Nutrients between the CNS and the Vascular System.
- Some molecules can cross the membrane while others cannot. This accounts for the inability of many drugs to cross the Blood-Brain Barrier.

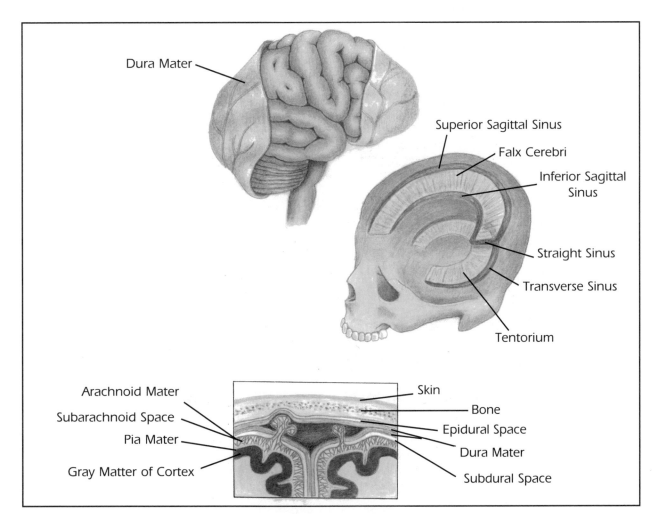

Spinal Cord Anatomy

SPINAL CORD ANATOMY

Boundaries of the Spinal Cord
- Boundaries of the SC extend from the FORAMEN MAGNUM to the CONUS MEDULLARUS.
- The CONUS MEDULLARIS is the END of the SC at the L1 – L2 vertebral area.
- The SC then becomes the CAUDA EQUINA (means "horse's tail").
- The CAUDA EQUINA are SPINAL NERVES that have not yet exited the Vertebral Column.

Enlargements of the Spinal Cord
- The SC has an Hourglass Shape. Enlargements occur in the CERVICAL and LUMBAR Sections.
- The Cervical Enlargement is due to the BRACHIAL PLEXUS—a network of spinal nerves from C5 – T1 that extend from the Cervical Vertebrae to the Upper Extremities.
  - When these Spinal Nerves enter the SC (and synapse with SC Tracts) they account for the large area of White Matter in the Cervical SC.
- The Lumbar Enlargement is due to the LUMBAR PLEXUS—a network of spinal nerves from L1 – S3 that extend from the Lumbar Vertebrae to the Lower Extremities.
  - When these Spinal Nerves enter the SC (and synapse with SC Tracts) they account for the large area of White Matter in the Lumbar SC.

Anterior Median Fissure
- Continues from the Anterior Aspect of the Medulla to the End of the SC.

Dorsal Median Sulcus
- Continues from the Posterior Aspect of the Medulla to the End of the SC.

Dorsal Intermediate Sulcus
- Continues from the Posterior Aspect of the Medulla and extends only throughout the THORACIC LEVELS of the SC.
- The Dorsal Intermediate Sulcus separates two Ascending Sensory tracts—the Dorsal Columns, where the Fasciculus Gracilis and Cuneatus are located.

Central Canal
- Contains cerebrospinal fluid (CSF).

SPINAL CORD ANATOMY: PNS VS. CNS

Spinal Nerves
- The Spinal Nerves are located in the PNS.
- Ascending Sensory Spinal Nerves extend from a Sensory Receptor to the Dorsal Rootlets.
- Descending Motor Spinal Nerves extend from the Ventral Horn of the SC to Skeletal Muscles.
- There are 31 pairs of Spinal Nerves:
  - 8 Cervical, 12 Thoracic, 5 Lumbar, 5 Sacral, and 1 Coxygeal

Dorsal Root Ganglion
- Contains the Cell Bodies of Sensory Nerves that are part of the somatic PNS.
- Each Sensory Nerve has its own Dorsal Root Ganglion.
- The Dorsal Root emerges from the Dorsal Ganglia.

Dorsal Root and Rootlets
- The Dorsal Roots are Ascending Spinal Nerves that carry Sensory Data from the Sensory Receptors (in the PNS) to the Dorsal Horn of the SC.
- Dorsal Roots are Axon Bundles that EMERGE from a Spinal Nerve.
- The Dorsal Root leads into the Dorsal Rootlets—thin string-like Axons that emerge from the Dorsal Root and Synapse in the Dorsal Horn of the SC.
- The Dorsal Root and Rootlets are considered to be within the PNS.

Dorsal Horn
- The Dorsal Horn is considered to be part of the CNS.
- The Dorsal Horn contains the Cell Bodies of many of the Sensory Spinal Cord Tracts.
- In the Dorsal Horn, the Dorsal Rootlets (of the PNS) may synapse on Interneurons. These Interneurons then synapse with SC Tracts. Or the Rootlets may synapse directly on the Cell Bodies of the SC Tracts.
- When the Spinal Nerves have Synapsed with a SC Tract, the SC TRACT Ascends through the SC, Brainstem, and travels to the Cortex.

Ventral Horn, Root, and Rootlets
- Descending Motor SC Tracts travel from the Cerebrum down through the Brainstem and SC.
- Motor SC Tracts synapse with Interneurons in the VENTRAL HORN.
- These Interneurons then synapse with Motor Spinal Nerves and exit the Ventral Horn through the Ventral Rootlets.
- The Ventral Rootlets merge into the Ventral Roots and extend to Skeletal Muscles.
- The VENTRAL HORN, ROOTLETS, AND ROOT are all considered to be within the PNS.

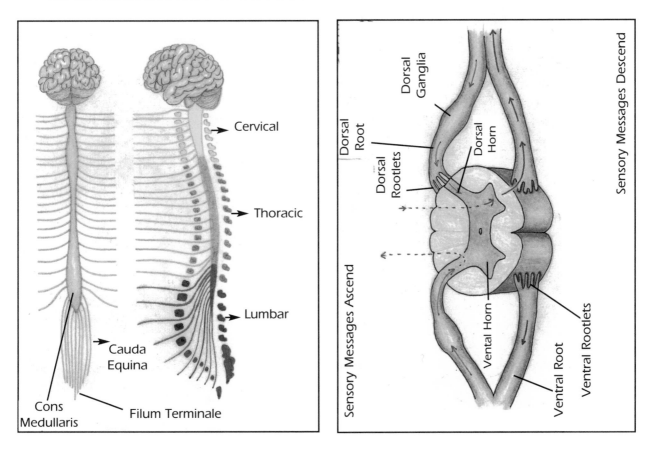

SPINAL NERVES AND THE DERMATOMES

Dermatome Distribution
- A DERMATOME is a Skin Segment that receives its innervation from a Specific Spinal Nerve.

Referred Pain
- Referred Pain occurs when a specific body region shares its spinal nerve innervation with a separate dermatomal skin segment.
- The pain experienced by the body part is misinterpreted by the Cortex as pain coming from a separate dermatomal skin segment.
- Example: Referred Pain in Heart Attack.
  - The Spinal Nerves that innervate the Heart also innervate the Left Arm (T1).
  - When the Heart experiences pain, the Cortex misinterprets the origin of the pain as coming from the Medial Aspect of the Left Arm.
  - This Cortical misinterpretation occurs because the Cortex does not have prior experience interpreting pain from the Heart. The Cortex relies on past experience when interpreting pain from a visceral source. Because it is uncommon for pain sensations to originate in the viscera, the Cortex initially interprets the pain as coming from the Left Arm (T1 Dermatome).
  - As the pain increases, the Cortex is able to correctly identify the source of the pain as the Heart.

Clinical Use of the Dermatomal Distribution
- When a therapist performs a sensory evaluation and determines that a specific body region does not register sensation, the therapist is then able to identify the lesion level.
- For example, if a patient cannot perceive sensation on the dorsal forearm, the therapist is able to determine that there is some impairment at C6 level.
- TENS UNIT (Transcutaneous Electrical Nerve Stimulation). The use of TENS is based on this principle.
  - The therapist places the TENS Unit on the identified Dermatome Region to Stimulate Nerve Regeneration or to Reduce Pain in a peripheral nerve injury.

SPINAL NERVES AND THE VERTEBRAL COLUMN

Relationship of the Spinal Cord to the Vertebral Column
Ontogenetic Development
- The SC is the same length as the Vertebral Column In Utero.
- But the Vertebral Column continues to grow after birth. The SC does NOT continue to grow.
- The Adult SC Ends at the L1 – L2 vertebral region.
- The remainder of the Spinal Nerves (the CAUDA EQUINA) must descend through the Vertebral Column to exit their Intervertebral Foramina.

There Is One More Pair of Spinal Nerves Than There Are Vertebrae
- From C1 – C7 the Spinal Nerves exit ABOVE their Corresponding Vertebrae.
- There is a pair of C8 Spinal Nerves, but NO C8 Vertebra.
- This means that the C8 Spinal Nerve must exit BELOW C7 Vertebra and ABOVE T1 Vertebra.
- T1 Spinal Nerve exits BELOW T1 Vertebra and ABOVE T2 Vertebra.
- From C8 down, the Spinal Nerves exit BELOW their corresponding Vertebrae.

Intervertebral Discs
- NUCLEUS PULPOSA is a soft, pulpy, highly elastic tissue in the center of the intervertebral disc.
- ANNULO FIBROSA is the more fibrous outer-covering of the disc.

Ruptured Disc
- The Nucleus Pulposa is the most likely part of the disc to rupture.
- When it ruptures, it travels to the place of least resistance—the Intervertebral Foramina.
- This results in a PINCHED SPINAL NERVE because the nerves exit the Vertebral Column through the Intervertebral Foramina.
- The Disc has HERNIATED.

Cervical Rupture

- Cervical Nerves exit through the Foramina ABOVE their corresponding vertebra.
- When a Cervical Disc has ruptured, the nerve ABOVE the rupture will be impinged.
- Example: A ruptured C5 disc will impinge C5 spinal nerve.

Lumbar Puncture

- The Lumbar and Sacral Nerves exit through the foramina BELOW their corresponding vertebrae.
- Because the CAUDA EQUINA forms the Lumbar Plexus, a ruptured Lumbar or Sacral Disc will often impinge SEVERAL Spinal Nerves.
- An example is SCIATICA—pain that radiates down the leg due to a ruptured disc in the Lumbar or Sacral regions. A ruptured disc in the Lumbar-Sacral regions will impinge several spinal nerves that innervate the lower extremities.

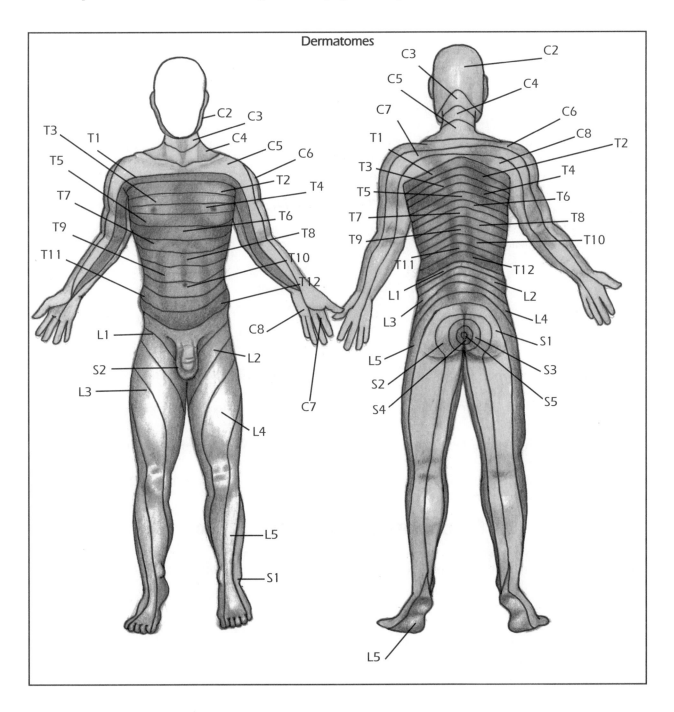

Dermatomes

CROSS-SECTIONS OF THE SPINAL CORD

Cervical Levels

- The Cervical Sections are LARGE and OVAL in appearance.
- Consist of a Large Amount of WHITE MATTER. These are the AXONS of the Sensory and Motor Tracts.
- The Descending Motor Tracts have not yet exited the SC.
- Most of the Ascending Sensory Tracts have already entered the cord.
- The amount of GRAY MATTER is Small.

Thoracic Levels

- Also OVAL shaped but are SMALLER than the Cervical Levels.
- The Thoracic Sections have a LATERAL HORN (also called an Intermediolateral Horn).
- The Lateral Horn is part of the AUTONOMIC NERVOUS SYSTEM (ANS).
- This is where the Cell Bodies for the SYMPATHETIC NERVOUS SYSTEM are located.

Lumbar Levels

- More ROUND in shape and LARGE.
- The Lumbar Sections have the LARGEST amount of GRAY MATTER.

Sacral Levels

- Look much like the Lumbar Sections but are much SMALLER.
- Most of the Descending Motor Tracts have already exited the SC.
- Many of the Ascending Sensory Tracts have not yet entered the SC.

ORGANIZATION OF THE INTERNAL SPINAL CORD

White Matter

- Consists of Myelinated Axons.
- The White Matter is divided into three pairs of FUNICULI:
  - Anterior
  - Lateral
  - Dorsal
- The SC TRACTS are located in the Funiculi.

Gray Matter

- Contains the Cell Bodies of the Sensory SC Tracts (in Dorsal Horn) and the Cell Bodies of the Motor Spinal Nerves (in Ventral Horn).
- The Cell Bodies for the Motor Spinal Nerves (that innervate the Skeletal Muscles) are organized in a precise pattern in the Ventral Horn.
  - The Cell Bodies for the Motor Spinal Nerves that innervate the PROXIMAL MUSCLE GROUPS are located in the Medial Ventral Horn.
  - The Cell Bodies for the Motor Spinal Nerves that innervate the DISTAL MUSCLE GROUPS are located in the Lateral Ventral Horn.

BLOOD SUPPLY OF THE SPINAL CORD

- The Blood Supply of the SC comes form the VERTEBRAL ARTERIES.
- The Vertebral Arteries are two Branches that give rise to one ANTERIOR SPINAL ARTERY and two POSTERIOR SPINAL ARTERIES.

Anterior Spinal Artery (1)

- The Vertebral A originates in the Medulla and sends off a branch called the Anterior Spinal Artery.
- The Anterior Spinal A descends down the Medulla and the Anterior Aspect of the SC.
- Runs along the Anterior Median Fissure.
- Supplies the Anterior Aspect of the SC.

Posterior Spinal Arteries (2)

- The Vertebral A also gives rise to two Posterior Spinal Arteries.
- These descend down the Dorsal Intermediate Sulci on the Posterior Aspect of the SC.
- Supply the Posterior Aspect of the SC.

Radicular Arteries

- The Radicular Arteries encircle the SC at all levels.
- The Radicular Arteries meet up with and supply the Anterior Spinal A and the Posterior Spinal As.

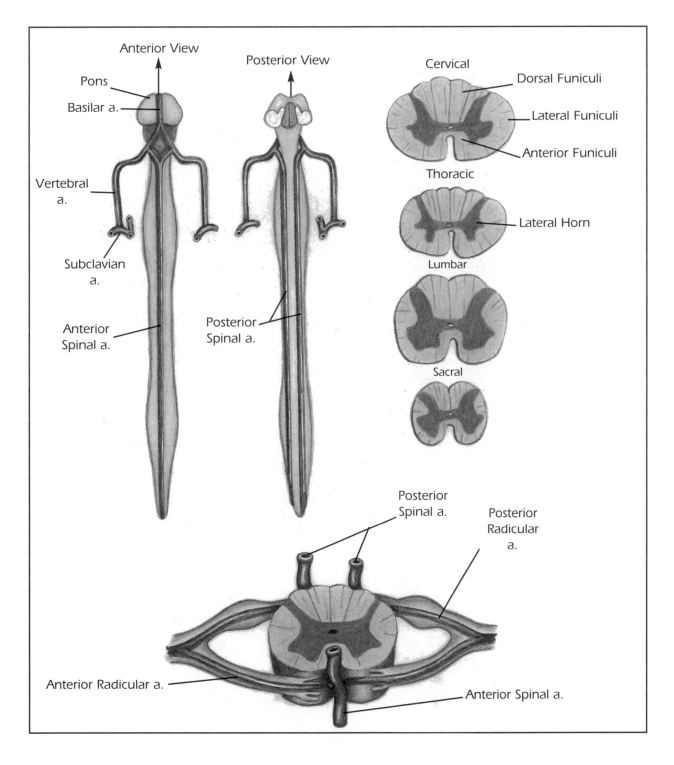

MENINGES OF THE SPINAL CORD

- The MENINGES of the SC are the same as those of the Brain.
- Function: to Protect and Anchor the SC.
- There are three Layers of Meninges.

Dura Mater

- The Dura is the Most Superficial and Thickest Membrane.

Arachnoid Mater

- The Middle Meningeal Membrane.
- CSF bathes the SC in the Subarachnoid Space.

Pia Mater

- Deepest and Thinnest Membrane.
- Adheres to the SC.
- Sends off two Projections: FILUM TERMINALE and DENTATE LIGAMENTS.

Filum Terminale (Projection of the Pia)

- The Filum Terminale is a slender median fibrous thread that attaches the Conus Medullaris to the Coccyx.
- Anchors the end of the SC to the Vertebral Column.

Dentate Ligaments (Also Called Denticulate Ligaments; Projection of the Pia)

- The Dentate Ligaments are a series of 22 triangular bodies that anchor the SC.

SPINAL REFLEX ARC

- A Spinal Reflex Arc is a reflex that is mediated at the Spinal Cord Level. There is no Cortical Involvement—no conscious decision-making.
- Pathway of a Spinal Reflex Arc:
  - A Sensory Receptor in the PNS sends a message along an Ascending Sensory Spinal Nerve.
  - The Sensory Spinal Nerve travels to the Dorsal Horn where it synapses on an Interneuron.
  - The Interneuron synapses on a Motor Cell Body located in the Ventral Horn.
  - The Motor Cell Body in the Ventral Horn relays the message to a Motor Spinal Nerve in the PNS.
  - The message travels to a Skeletal Muscle Group for action in response to the initial sensory message.

DEEP TENDON REFLEXES

- Reflex Arc in which a muscle contracts when its tendon is percussed.
- Also called Myotatic Reflexes, Monosynaptic Reflexes, and Muscle Stretch Reflexes.
- Deep Tendon Reflexes work on the principle of the Spinal Reflex Arc.
- Common Deep Tendon Reflexes are:
  - Biceps, Brachioradialis, Triceps, Patella (Knee), and Achilles Tendon (Ankle).
- In an Upper Motor Neuron Injury, Deep Tendon Reflexes become HYPERREFLEXIVE.
- In a Lower Motor Neuron Injury, Deep Tendon Reflexes become HYPOREFLEXIVE.

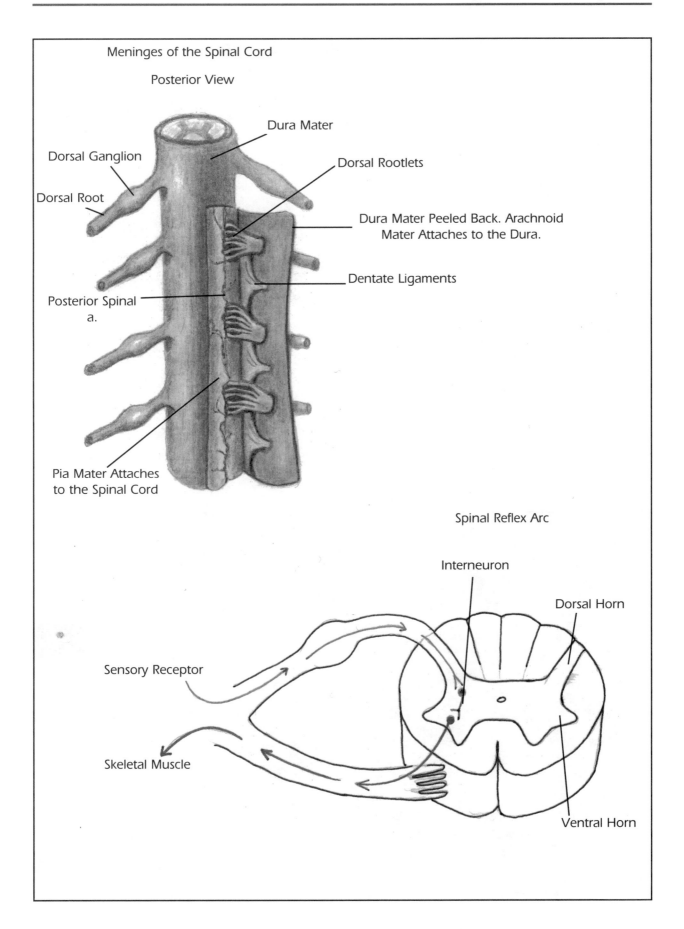

Meninges of the Spinal Cord

Posterior View

Dura Mater

Dorsal Ganglion

Dorsal Rootlets

Dorsal Root

Dura Mater Peeled Back. Arachnoid Mater Attaches to the Dura.

Dentate Ligaments

Posterior Spinal a.

Pia Mater Attaches to the Spinal Cord

Spinal Reflex Arc

Interneuron

Dorsal Horn

Sensory Receptor

Skeletal Muscle

Ventral Horn

UPPER VS LOWER MOTOR NEURONS

- Motor Neurons carry Motor Messages from different areas of the nervous system.
- Motor Neurons are divided into two Categories: Upper and Lower Motor Neurons.
- An UPPER MOTOR NEURON (UMN) carries motor messages from the Primary Motor Cortex to:
  - The Cranial Nerve Nuclei (located in the Brainstem).
  - Interneurons in the Ventral Horn. An UMN travels up to, but does not actually enter, the Ventral Horn.
  - An UMN is considered to be part of the CNS.
- A LOWER MOTOR NEURON (LMN) carries motor messages from the Motor Cell Bodies in the Ventral Horn to the Skeletal Muscles in the periphery.
  - A LMN is considered to be part of the PNS.
  - LMNs include the Cranial Nerves, Spinal Nerves, Cauda Equina, and the Ventral Horn.
- In an UMN Lesion, SPASTICITY occurs BELOW the Lesion Level.
  - Spasticity occurs because the Spinal Reflex Arcs Below the lesion level remain intact.
  - The Spinal Reflex Arcs operate without Cortical Modification.
  - Thus, increased muscle tone (or spasticity) occurs.
- In an UMN Lesion, FLACCIDITY occurs AT the Lesion Level.
  - Flaccidity occurs because the Spinal Reflex Arc AT the Lesion Level is lost.
  - Thus, nothing is innervating the muscles.
  - The muscles (at the lesion level) lose all tone and become flaccid.
- In a LMN Lesion, FLACCIDITY occurs AT and BELOW the Lesion Level.
  - Flaccidity occurs in all LMN Lesions because a LMN does not involve any Spinal Reflex Arcs.
  - Because Spinal Reflex Arcs are not part of LMN Lesions, there is nothing that continues to innervate the muscles.

Congenital Anomilies of the Spinal Cord

Spina Bifida

- Spina Bifida is a bony defect; it is the incomplete closure of the Vertebral Column during Fetal Development. A section of the SC does not unite in midline. Usually the opening occurs in the low Thoracic and Lumbar Sections of the SC.
- Meningocele
  - A Meningocele occurs when the Meninges and the CSF protrude through the opening of the vertebral column.
  - This results in Compression of the SC and some Nerve Roots.
  - There is a visible Cyst on the infant's back that is filled with CSF and Neural Tissue.
- Meningomyelocele
  - This is a severe form of Spina Bifida.
  - Occurs when the Meninges, SC, and Spinal Nerves all protrude.
  - All Motor and Sensory Information Below the Level of the Cyst are lost.

Arnold Chiari Malformation

- Involves displacement of the Cerebellar Vermis, Brainstem, and the 4th Ventricle.
- The Cerebellar Tonsils are often displaced downward into the Upper Cervical Canal.
- Arnold Chiari Malformation is often associated with a Myelomeningocele.
- Individuals with Arnold Chiari Malformation may present with Hydrocephalus and Cerebellar and Lower Cranial Nerve Signs.

Spina Bifida

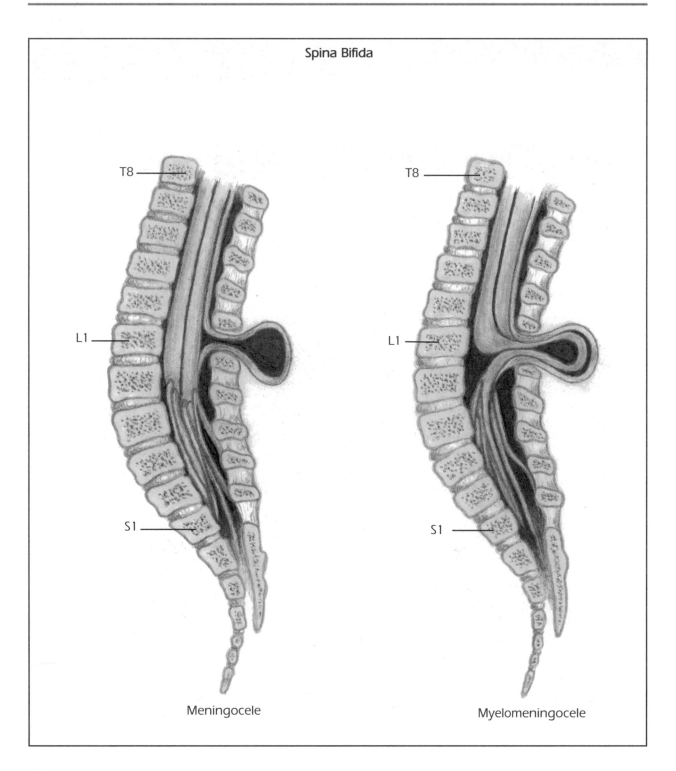

T8

L1

S1

Meningocele

T8

L1

S1

Myelomeningocele

The Cranial Nerves

THE CRANIAL NERVES

Cranial Nerve Anatomy
- There are 12 pairs of Cranial Nerves (CNs).
- The CNs are considered to be part of the Peripheral Nervous System (PNS).
- Their Nuclei (Cell Bodies) are Located in the Brainstem.
- CN Nuclei are considered to be part of the Central Nervous System (CNS).
- CNs begin exiting the Brain at the Midbrain Level and lead all the way down the Medulla.
- CNs use Roman Numerals (however, Arabic Numbers will be used in this text).

List of Cranial Nerves
1. OLFACTORY N
2. OPTIC N
3. OCULOMOTOR N
4. TROCHLEAR N
5. TRIGEMINAL N
6. ABDUCENS N
7. FACIAL N
8. VESTIBULOCOCHLEAR N
9. GLOSSOPHARYNGEAL N
10. VAGUS N
11. ACCESSORY N
12. HYPOGLOSSAL N

Function
- CNs carry SENSORY and MOTOR Information to and from the following Receptors of the HEAD, FACE, and NECK:
  - SPECIAL SENSE Receptors (for Vision, Audition, Olfaction, Gustation, Equilibrium)
  - SOMATOSENSORY Receptors
  - PROPRIOCEPTORS

Most CN Lesions Produce IPSILATERAL Signs and Symptoms.

Cranial Nerves

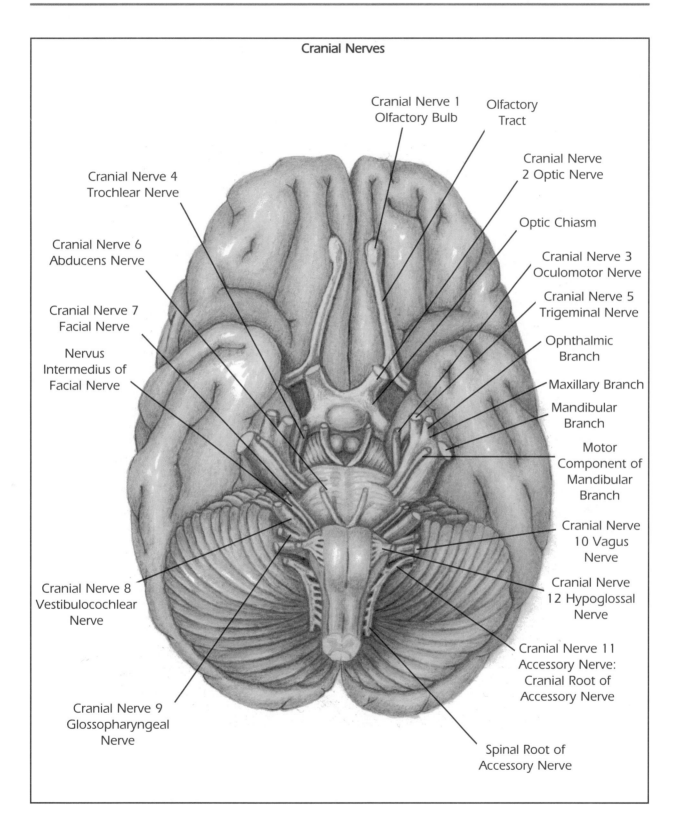

Cranial Nerve 1
Olfactory Bulb

Olfactory
Tract

Cranial Nerve
2 Optic Nerve

Optic Chiasm

Cranial Nerve 3
Oculomotor Nerve

Cranial Nerve 5
Trigeminal Nerve

Ophthalmic
Branch

Maxillary Branch

Mandibular
Branch

Motor
Component of
Mandibular
Branch

Cranial Nerve
10 Vagus
Nerve

Cranial Nerve
12 Hypoglossal
Nerve

Cranial Nerve 11
Accessory Nerve:
Cranial Root of
Accessory Nerve

Spinal Root of
Accessory Nerve

Cranial Nerve 4
Trochlear Nerve

Cranial Nerve 6
Abducens Nerve

Cranial Nerve 7
Facial Nerve

Nervus
Intermedius of
Facial Nerve

Cranial Nerve 8
Vestibulocochlear
Nerve

Cranial Nerve 9
Glossopharyngeal
Nerve

CN 1: OLFACTORY NERVE

Carries
- SENSORY Information.

Nuclei Location
- Nasal Receptors in the Nose.

Function
- Olfaction (smell).

Pathway
- Nasal Receptors in the Nose send Olfactory Messages to the Inferior Frontal Lobes—where the Olfactory Bulb is Located.
- The Olfactory Information then travels from the Olfactory Bulb down the Olfactory CN to the Hippocampal Formation in the Temporal Lobe.

Lesion Symptoms
- If the Lesion is Unilateral (only occurs in one Olfactory Nerve): There are NO Symptoms—because the opposite Olfactory Nerve Compensates for the Lost Sense of Smell on one side.
- If the Lesion is Bilateral: The individual Loses the Sense of Smell.
- ANOSMIA
- Lack of Olfactory Function.
- Often, Anosmia occurs as a result of Head Injury.

Test
- Test one Olfactory Nerve at a Time.
- Occlude Vision.
- Block the patient's nostril on the opposite side being tested.
- Present one odor at a time.
- Provide the patient with a verbal choice of specific odors if he or she has difficulty identifying the odors but can easily smell them or if the patient has word finding difficulties.

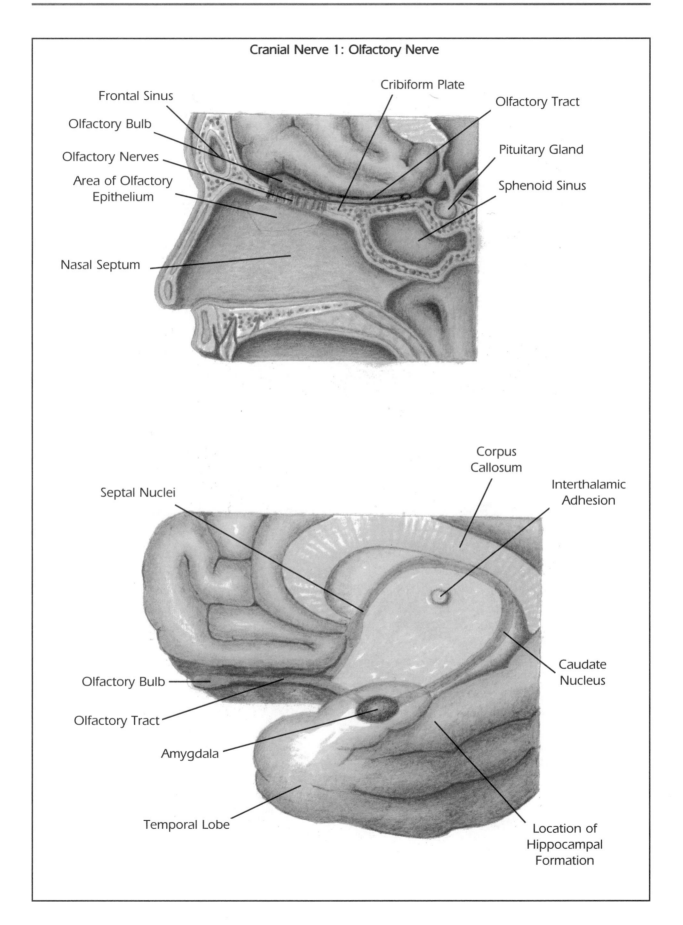

Cranial Nerve 1: Olfactory Nerve

CN 2: Optic Nerve

Carries
- SENSORY Information.

Nuclei Location
- Visual Receptors of the Retina (Rods and Cones).

Function
- Visual Acuity (the ACCURACY of SIGHT; NOT the Interpretation of Visual Information).
- Visual Messages that travel from the Thalamus to the Superior Colliculi (and back) are involved in:
    - Pupillary Reflexes
    - Awareness of Light and Dark
    - Orientation of Head and Eye Movements

Pathway
- The Rods and Cones of the Retina send Visual Information down the OPTIC NERVES to the OPTIC CHIASM.
- The Visual Information then travels from the OPTIC CHIASM through the OPTIC TRACTS.
- Visual Information travels from the OPTIC TRACTS to the LATERAL GENICULATE BODIES of the Thalamus.
- A branch carries Visual Messages from the LATERAL GENICULATE BODIES of the Thalamus to the SUPERIOR COLLICULI of the Midbrain (this information is then processed back through the Thalamus).
- Visual Messages then travel to the OCCIPITAL LOBES for Visual Detection and Interpretation.

Lesion Symptoms
- A Unilateral Lesion (in only one OPTIC N) produces IPSILATERAL BLINDNESS.
- A Bilateral Lesion (in both OPTIC Ns) produces BILATERAL BLINDNESS.

Test
- When Testing OPTIC N Function, three kinds of data should be collected (or the Examination is not complete).
    - Results from a VISUAL ACUITY TEST (SNELLEN EYE CHART)
    - Results from a VISUAL FIELD TEST
    - Results from a FUNDOSCOPIC EXAM
- Generally, therapists perform Visual Acuity Tests and Visual Field Tests. Ophthalmologists perform Fundoscopic Exams.

Visual Acuity Test (Snellen Eye Chart)
- Test one Eye at a time. Then test both eyes together.
- If the patient normally wears corrective lenses, test the patient with glasses on.

Visual Field Test
- Test one Eye at a time. Occlude vision in opposite eye.
- Test Vertical and Temporal Peripheral Vision.
- Normal Vertical Vision is 45 degrees.
- Normal Temporal Vision is 85 degrees.

Cranial Nerve 2: Optic Nerve

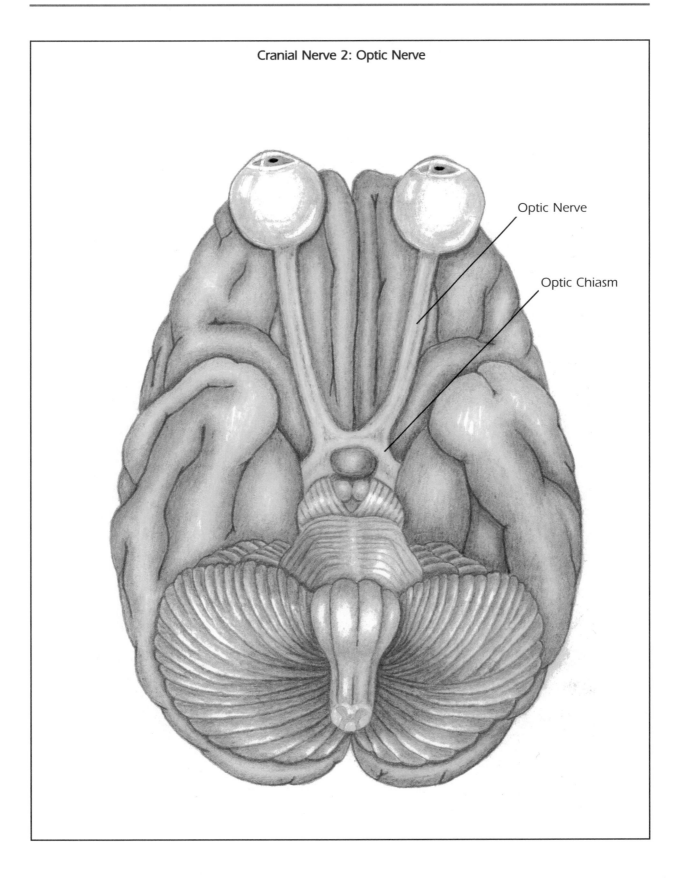

Optic Nerve

Optic Chiasm

CN 3: Oculomotor Nerve

Carries

- MOTOR Information.

Nuclei Location

- Midbrain at the Level of the Superior Colliculi.

Function

- Extraocular Eye Movements.
- CN 3 is responsible for the eyeball movements up, down, medially, and laterally.
- The OCULOMOTOR N innervates the eye muscles that control these movements.
- The Oculomotor N is considered to be one of the EXTRAOCULOMOTOR NERVES, along with the Trochlear and Abducens Ns.
- The three Extraoculomotor Ns use the MEDIAL LONGITUDINAL FASCICULUS to communicate with each other and with the Vestibular System.
- The Medial Longitudinal Fasciculus is a Brainstem Tract that coordinates HEAD and EYE MOVEMENTS by providing bilateral connections among the Vestibular Nerve Nuclei, Extraocular Nerve Nuclei, and the Accessory Nerve Nuclei in the Brainstem.

The Oculomotor Nerve Innervation

1. LEVATOR PALPEBRAE: Responsible for Lifting the Eyelid.
2. SUPERIOR RECTUS MUSCLE: Responsible for Upward/Medial eyeball movements.
3. INFERIOR OBLIQUE M: Responsible for Upward/Lateral eyeball movements.
4. INFERIOR RECTUS M: Responsible for Downward/Medial eyeball movements.
5. MEDIAL RECTUS M: Responsible for Medial eyeball movements.

Inferior Oblique		Superior Rectus	
(CN 6) Lateral Rectus		Medial Rectus	NOSE
Superior Oblique (CN4)		Inferior Rectus	

Oculomotor Nerve Reflexes

1. PUPILLARY REFLEX: Pupil of the eye constricts when light is shined into it.
2. ACCOMMODATION: Lens of the eye adjusts to focus light on the retina.
3. CONVERGENCE: Pupils move medially when viewing an object at close range.

Cranial Nerve 3: Oculomotor Nerve

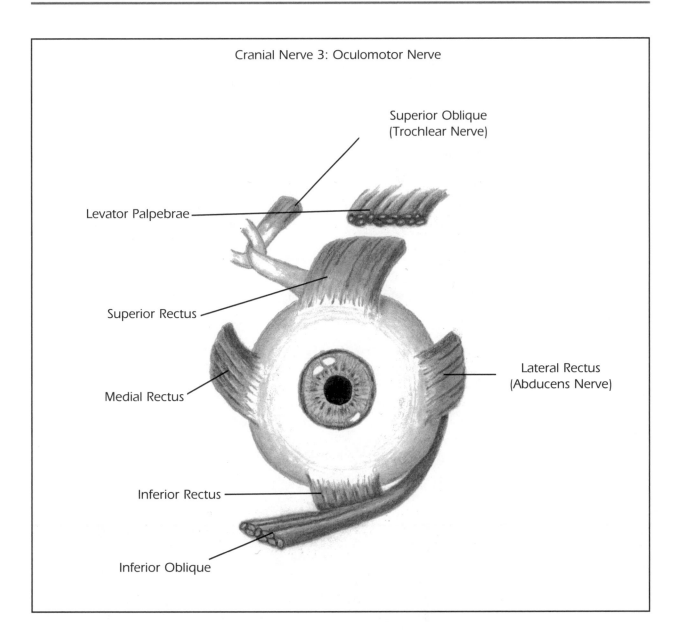

Superior Oblique
(Trochlear Nerve)

Levator Palpebrae

Superior Rectus

Medial Rectus

Lateral Rectus
(Abducens Nerve)

Inferior Rectus

Inferior Oblique

Lesion Symptoms

Lateral Strabismus (External Strabismus or Exotropia)

- The eyeball deviates outward (or Laterally) because the Medial Rectus is lost and the Lateral Rectus is working unopposed.
- Can cause DIPLOPIA, or double vision.

Ptosis

- Ptosis is the drooping of a body region; in this case the ipsilateral eyelid droops.

Nystagmus

- Involuntary back and forth movements of the eye in a quick, jerky, oscillating fashion when the eye moves laterally or medially to either the Temporal or Nasal extremes.
- Nystagmus can be normally elicited in an intact CNS using Rotational or Temperature Stimulation of the Semicircular Canals.
- Pathological Nystagmus is a sign of CNS abnormality and can occur with or without external stimulation.

Test

- Therapists test the EXTRAOCULOMOTOR Ns simultaneously—the OCULOMOTOR, TROCHLEAR , and the ABDUCENS Ns.
- Test one eye at a time. Occlude the eye not being tested.
- Instruct the patient to maintain head in a fixed position while visually scanning a moving stimulus.
- The moving stimulus can be the colored pen cap.
- The Therapist moves the Visual stimulus in the shape of an H (as shown above). This allows the Therapist to determine if the eyeball muscles are functioning adequately (ie, if the EXTRAOCULAR MUSCLES are adequately innervated by their EXTRAOCULOMOTOR CNs).
- Observe symmetry of pupil size.
- In dimly lit area, shine pen light at the bridge of the nose and observe for symmetrical corneal reflection. If corneal reflection is asymmetrical, strabismus is indicated.
- Shine pen light into one eye at a time for a 2-second duration. Check for Constriction of stimulated pupil.

Horner's Syndrome

- Can occur as a result of OCULOMOTOR N DAMAGE.
- Results from a lesion anywhere along the Sympathetic Pathway to the Pupil.
- Characterized by an IPSILATERAL SMALL PUPIL, IPSILATERAL MILD PTOSIS, and IPSILATERAL ANHIDROSIS (absence of sweating on the ipsilateral face and neck).
- Most common cause is a tumor involving the sympathetic fibers that lead to the pupil.

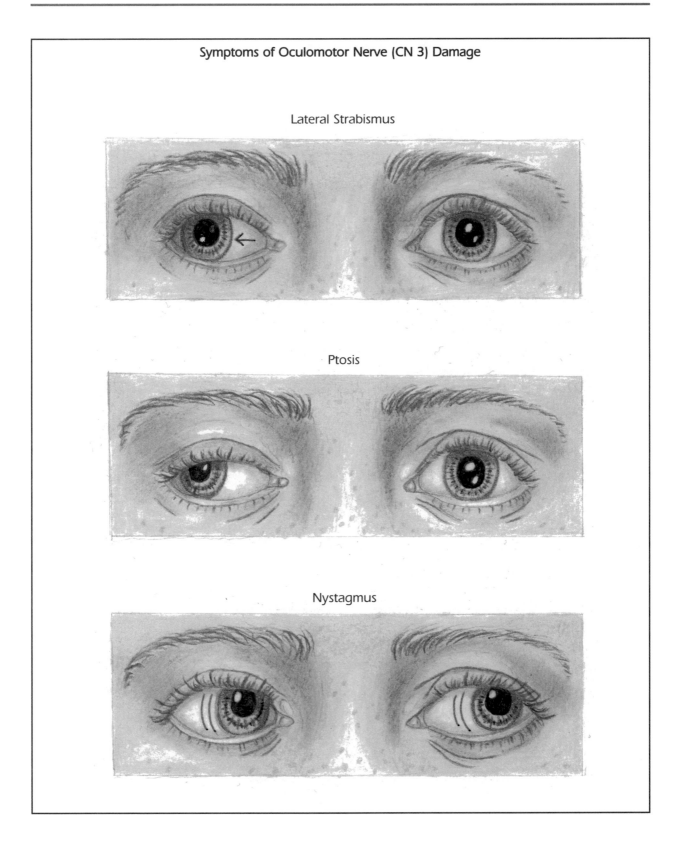

Symptoms of Oculomotor Nerve (CN 3) Damage

Lateral Strabismus

Ptosis

Nystagmus

CN 4: Trochlear Nerve

Carries
- MOTOR Information.

Nuclei Location
- Midbrain at the level of the Inferior Colliculi.

Function
- Extraocular Eye Movements.
- The TROCHLEAR N is responsible for Downward and Lateral eyeball movements.
- Considered to be one of the EXTRAOCULOMOTOR NERVES.

Innervates
- Superior Oblique Muscles.

Lesion Symptoms
- The patient will experience difficulty moving the eyeball DOWN and LATERALLY.
- This occurs because the Superior Oblique Muscle is lost. The Medial Rectus and the Superior Rectus Muscles are working unopposed to pull the eyeball up and medially.
- Results in a subtle VERTICAL, MEDIAL STRABISMUS.
- The patient may display difficulty walking down steps.
- VERTICAL DIPLOPIA is often reported at both near and far distances.
- NYSTAGMUS.

Test
- Test the Trochlear Nerve Function simultaneously with the Oculomotor and Abducens Ns (see Oculomotor N Testing).

Symptoms of Trochlear Nerve Damage (CN 4)

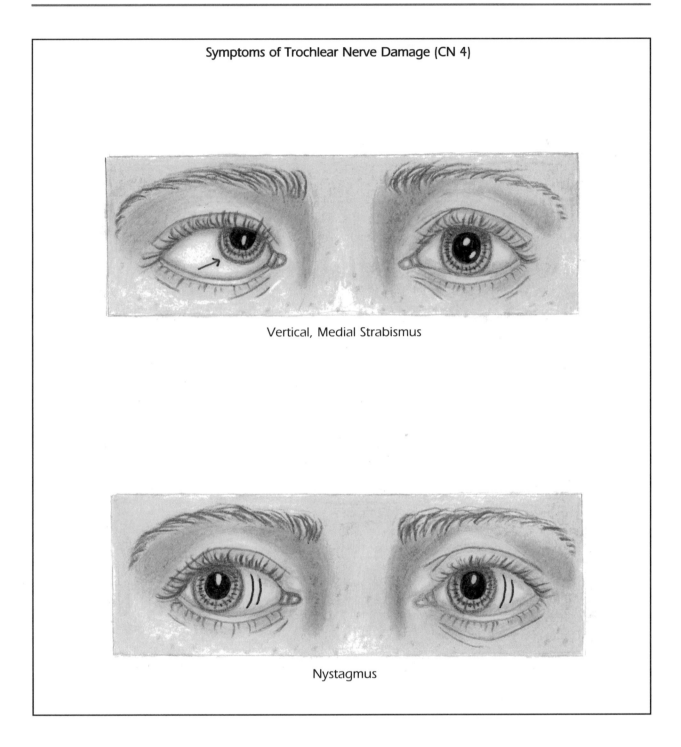

Vertical, Medial Strabismus

Nystagmus

CN 5: TRIGEMINAL NERVE

Carries
- SENSORY and MOTOR Information.

Nuclei Location
- Mid Pons.

Function

Sensory
- The Sensory Half of the Trigeminal N mediates SENSATION of the FACE, HEAD, CORNEA of EYE, and the inner ORAL CAVITY (the Ophthalmic, Maxillary, and Mandibular Regions).
- Sensation includes pain, temperature, and discriminative touch.

Motor
- The Motor Half of the Trigeminal N innervates the JAW MUSCLES that control CHEWING (or mastication).

Lesion Symptoms

Sensory
- Damage to the Sensory Half of the Trigeminal N causes Ipsilateral Loss of Sensation to the Head, Face, and Inner Oral Cavity.
- TRIGEMINAL NEURALGIA: occurs when one half of the face loses sensation.

Motor
- Damage to the Motor Half of the Trigeminal N causes weakness in chewing (mastication).
- The Jaw also deviates to the affected side.

Test

To Test the Sensory Half of the Trigeminal Nerve
- Evaluate the patient's sensory abilities on the Face, Head, and Inner Oral Cavity.
- Occlude vision.
- Use a cotton swab to stroke the inner oral cavity. Assess intact side first. Then evaluate involved side.
- Use a cotton swab to stroke the patient's forehead, cheek, jaw, and chin. Apply stimulus to the unaffected side first. Then proceed to involved side.
- Touch the patient's Cornea lightly with a cotton swab to check for Corneal Reflex—eyelid should close.

To Test the Motor Half of the Trigeminal Nerve
- Occlude vision.
- Ask patient to open his or her mouth. Check for deviation of the jaw to the affected side.
- Check for asymmetry of the size of the mouth opening—the patient will likely exhibit a decreased ability to open the mouth on the affected side.
- Ask the patient to move jaw from side to side. Check for asymmetry of jaw movement.
- Instruct the patient to bite down on a tongue depressor. Ask the patient to resist attempts, made by the therapist, to pull the tongue depressor out. Check for asymmetry between right and left jaw strength.

Trigeminal Nerve Reflexes

Masseter Reflex
- When the Masseter M is lightly tapped with a reflex hammer, the Masseter Contracts.

Corneal Reflex (Blink Reflex)
- When the cornea is touched, the eyelids close.

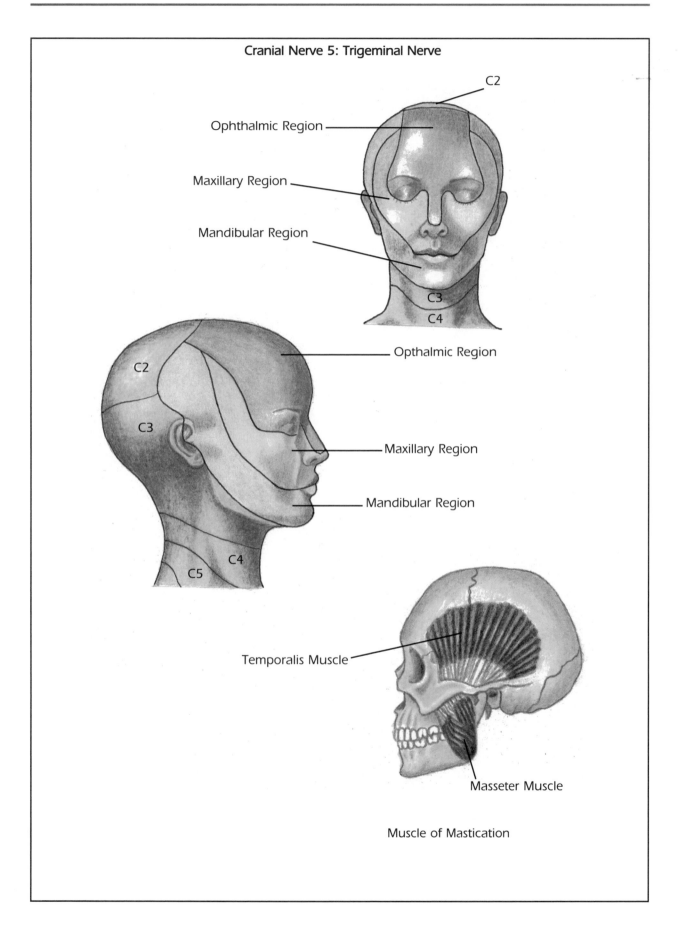

Cranial Nerve 5: Trigeminal Nerve

C2

Ophthalmic Region

Maxillary Region

Mandibular Region

C3

C4

Opthalmic Region

C2

C3

Maxillary Region

Mandibular Region

C4

C5

Temporalis Muscle

Masseter Muscle

Muscle of Mastication

CN 6: ABDUCENS NERVE

Carries
- MOTOR Information.

Nuclei Location
- Low Pons.

Function
- Extraocular Eye Movements.
- The Abducens N is responsible for Lateral Deviation of the eyeball—looking laterally.

Innervates
- Lateral Rectus Muscle.

Lesion Symptoms

Medial Strabismus (Internal Strabismus or Esotropia)
- Turning inward of the eyeball.
- This occurs because the Lateral Rectus M is lost. The Medial Rectus works unopposed to pull the eyeball medially.
- Can cause DOUBLE VISION, or DIPLOPIA.
- NYSTAGMUS.

Test
- The Abducens N is simultaneously tested with the Oculomotor and Trochlear Ns (see Oculomotor N Testing).

Symptoms of Abducens Nerve Damage (CN 6)

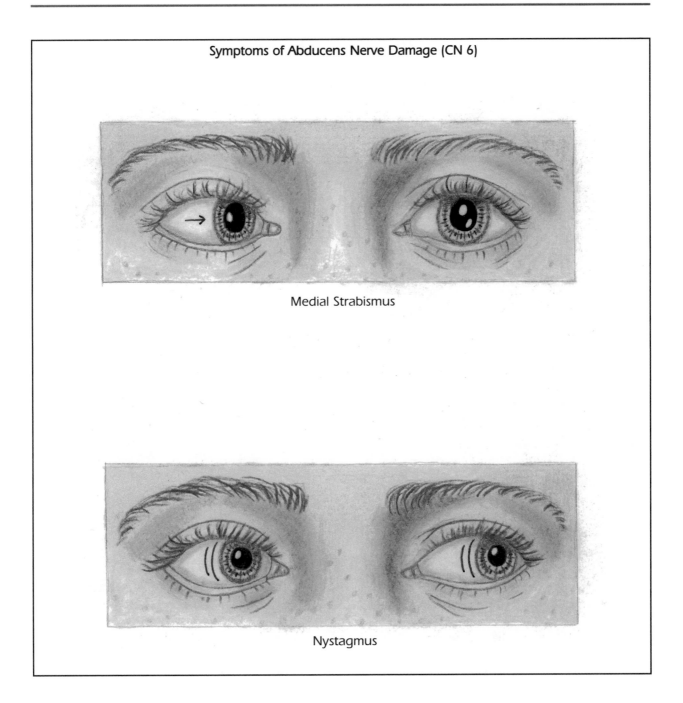

Medial Strabismus

Nystagmus

CN 7: Facial Nerve

Carries
- SENSORY and MOTOR Information.

Nuclei Location
- Mid to Low Pons.

Function

Sensory
- The Sensory Portion of the Facial N comes off of a separate branch of the Facial N called the NERVOUS INTER-MEDIUS.
- Innervates the TASTE RECEPTORS on the ANTERIOR TONGUE.

Motor
- The Motor Portion of the Facial N innervates the:
  - MUSCLES OF FACIAL EXPRESSION
  - MUSCLES FOR EYELID CLOSING
  - STAPEDIUS MUSCLE (controls the Stapes of the middle ear)

Lesion Symptoms
- Decreased Taste on the Anterior of the Tongue.
- Decreased Corneal Reflex.

Test

To Test the Sensory Portion of the Facial Nerve
- Test the sense of Taste on the Anterior Tongue.
- Occlude vision.
- Present SWEET, SALTY, and SOUR solutions to the outer and lateral portions of the anterior tongue.
- Present each taste substance one at a time. Ask the patient to indicate if he or she can taste the substance and identify whether it is sweet, sour, or salty.

To Test the Motor Portion of the Facial Nerve
- Test the STRENGTH and SYMMETRY of Facial Muscles.
- Ask patient to Elevate Eyebrows and Forehead.
- The ability to wrinkle the Forehead is used to distinguish an Upper Motor Neuron (UMN) Lesion from a Lower Motor Neuron (LMN) Lesion.
- In an UMN Lesion, the Muscles of the Forehead will be spared (remain intact), even while the lower facial regions are not. This commonly occurs in Cerebrovascular Accident.
- Bell's Palsy is a LMN disorder (The Facial CN has been lesioned). Both the Forehead and the Lower Face are involved in the paralysis of facial muscles.
- Ask the patient to smile, frown, and pucker lips. Check for asymmetry on the right and left sides of the face.
- Ask the patient to blow cheeks up with air. Gently push on cheeks while asking the patient to resist. Check for asymmetry of facial muscle strength.

Facial Nerve Reflex

- CORNEAL OR BLINK REFLEX: When the Cornea is touched, the eyelids close.

Bell's Palsy

- Drooping of the ipsilateral side of Face.
- The Etiology of Bell's Palsy is obscure. May be caused by a Viral Infection.
- Acute Phase is characterized by marked Edema causing tension to the Facial N.
- An Inflammatory Process begins. This induces Edema with secondary Vascular Compromise.
- Inflammation and Vascular Compromise cause anoxia to the Facial N.
- This leads to Vasodilation, transudation of fluid (oozing of fluid through pores), and further pressure that confines the pathway of CN 7.

Hyperacusis in Bell's Palsy
(increased sensitivity to sound)

- Because the Facial N travels through the Internal Auditory Meatus and later gives off branches to the Stapedius M, patient's with Bell's Palsy may complain of ipsilateral Hyperacusis.

CN 8: Vestibulocochlear Nerve

Carries
- Two SENSORY BRANCHES.

Nuclei Location
- Pons-Medulla Junction.

Function

Auditory Branch
- The Auditory Branch of the Vestibulocochlear N transmits sensory impulses that result from the vibrations of the fluid in the Cochlea.
- The Function of the Auditory Branch is Audition (hearing).

Vestibular Branch
- The Vestibular Branch receives sensory stimulation from the SEMICIRCULAR CANALS of the inner ear.
- Concerned with BALANCE and the sensations of VERTIGO (dizziness).
- The Function of the Vestibular Branch is BALANCE, EQUILIBRIUM, and the POSITION OF THE HEAD IN SPACE.

Pathway

Auditory Branch
- The Auditory Branch of the Vestibulocochlear N runs from the Hair Cells (or receptors) of the Organ of Corti (in the inner ear) to the Vestibular Nucleus in the Brainstem.

Vestibular Branch
- The Vestibular Branch runs from the SEMICIRCULAR CANALS, UTRICLES, and SACCULES (of the inner ear) to the Vestibular Nuclei in the Brainstem.
- The Semicircular Canals, Utricles, and Saccules detect changes in Head Position.

Lesion Symptoms

Auditory Branch Lesions
- Deafness or Tinnitus.

Vestibular Branch Lesions
- Nystagmus (due to connections to the Extraoculomotor Nerve Nuclei).
- Vertigo.
- Decreased Balance.
- Decreased Protective Responses.
- Changes in Extensor Tone (because the Vestibulospinal Tract is responsible for mediating Extensor Tone).

Test

To Test Auditory Branch of the Vestibulocochlear Nerve

An Audiologist Must First Distinguish Between Two Possible Types of Hearing Impairment:
- SENSORINEURAL: Involves the Inner Ear, Vestibulocochlear N, and the Brain.
- CONDUCTIVE: Involves the Outer Ear and the Middle Ear Structures.

To Test the Vestibular Branch of the Vestibulo-cochlear Nerve
- TEST FOR NYSTAGMUS.
  - Have patient in seated position.
  - Have patient track a moving object (at a distance of 15 inches) in an H and X pattern.
  - Check for Nystagmus both within and at end ranges of visual fields.
- TEST BALANCE AND PRESENCE OF PROTECTIVE RESPONSES (Romberg Test).
  - Have patient stand with eyes open, then closed.
  - Check for increased sway and loss of balance.
  - Gently displace patient's balance. Check for Protective Responses.
- TEST FOR PRESENCE OF EXTENSOR TONE IN LOWER EXTREMITIES.

Vestibular Neuritis

- Inflammation of the Vestibular Nerve, usually caused by a virus.
- Results in Dysequilibrium, Nystagmus, Nausea, and Severe Vertigo that can persist for several days.

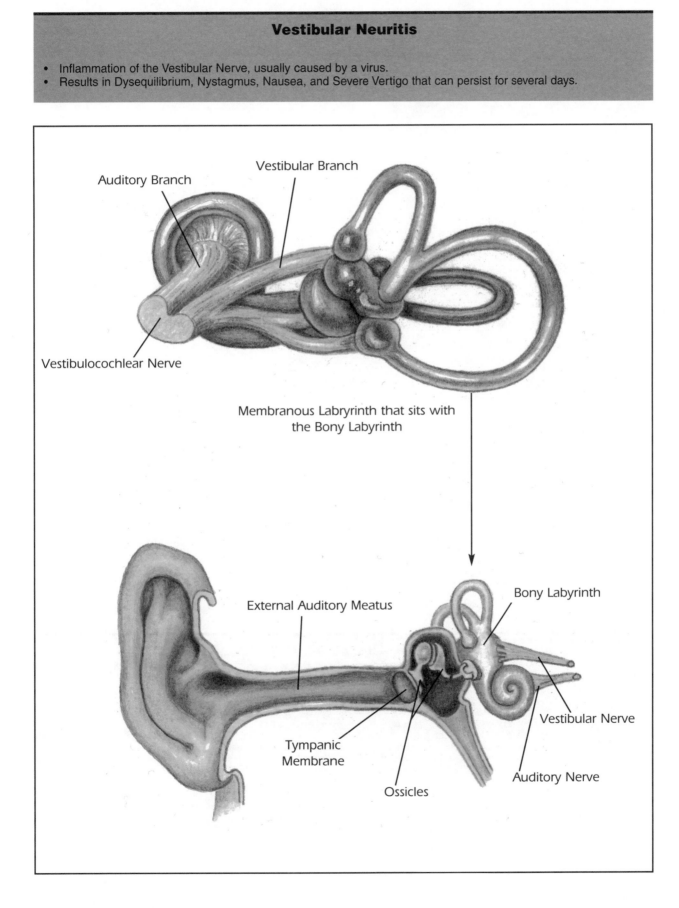

Auditory Branch

Vestibular Branch

Vestibulocochlear Nerve

Membranous Labryrinth that sits with the Bony Labyrinth

Bony Labyrinth

External Auditory Meatus

Vestibular Nerve

Auditory Nerve

Tympanic Membrane

Ossicles

CN 9: Glossopharyngeal Nerve

Carries
- SENSORY and MOTOR Information.

Nuclei Location
- Nucleus Ambiguous in the Medulla.

Function

Sensory
- Taste on the Posterior Aspect of the Tongue.

Motor
- Swallowing.

Lesion Symptoms

Sensory
- Loss of Taste Sensation on Posterior Aspect of Tongue (Loss of Bitter Taste Modality).

Motor
- Loss of the Gag and Swallowing Reflexes.
- DYSPHAGIA: Difficulty Swallowing, leading to Choking or Food Aspiration.

Test
- Because they mediate similar functions, the GLOSSOPHARYNGEAL and VAGUS Ns are Tested Simultaneously.

Sensory
- Use the same testing procedures described for the Sensory Half of the Facial Nerve, except apply the procedure to the Posterior Aspect of the Tongue (where Bitter Tastes are detected).
- Ask patient to chew on a lemon rind to determine if bitter tastes can be detected.

Motor
- Attempt to elicit the Gag Reflex by swiping a tongue depressor or cotton swab at the back of the throat.
- Observe the patient's ability to swallow different consistencies of food.
- Present different consistencies of food one at a time (solid foods, pureed foods, thick liquids, thin liquids).
- Ask patient to consume each presented food type. Check for food aspiration, coughing, throat clearing, or a wet vocal quality (indicating that the food is pocketing in the larynx).
- If Aspiration Precautions have been indicated previously, modify the above procedure accordingly.

GLOSSOPHARYNGEAL NERVE REFLEXES

Gag Reflex
Touching of the Pharynx elicits contraction of the Pharyngeal Muscles.

Swallowing Reflex
Food touching the Pharynx elicits Movement of the Soft Palate and Contraction of the Pharyngeal Muscles.

CN 10: Vagus Nerve

Carries
- SENSORY and MOTOR Information.

Nuclei Location
- 1. Dorsal Vagal Nuclei in the Medulla.
- 2. Nucleus Ambiguous in the Medulla.

Function

Visceral Branches (to Hollow Organs)
- Carry SENSORY and MOTOR Information.
- Carry Parasympathetic Information to and from the HEART, PULMONARY SYSTEM, ESOPHAGUS, and the GASTROINTESTINAL TRACT.

Skeletal Muscle Branches
- Carry MOTOR Information to the muscles of the LARYNX, PHARYNX, and UPPER ESOPHAGUS.
- These Muscles are responsible for Swallowing and Speaking.

Lesion Symptoms

Visceral Branch Lesions
- Transient Tachycardia (irregular rapid heart beat).
- DYPSNEA: Difficulty Breathing.

Bilateral Visceral Branch Lesions
- ASPHYXIA (suffocation).

Skeletal Muscle Branch Lesions
- DYSPHONIA (hoarse voice).
- DYSPHAGIA (difficulty swallowing).
- DYSARTHRIA (difficulty articulating words clearly—slurring words).

Test
- The GLOSSOPHARYNGEAL and VAGUS Ns are tested simultaneously. See testing procedures under Glossopharyngeal Nerve.
- Also observe the patient's ability to speak clearly without slurring words.
- Check for decreased phonal volume, hoarse voice.

Vagus Nerve Reflexes

Gag Reflex
Touching of the Pharynx elicits Contraction of the Pharyngeal Muscles.

Swallowing Reflex
Food touching the Pharynx elicits Movement of the Soft Palate and Contraction of the Pharyngeal Muscles.

CN 11: ACCESSORY NERVE

Carries

- MOTOR Information.

The Accessory Root Has Two Nerve Roots

Cranial Nerve Root

- Emerges from the Nucleus Ambiguous and Joins the Vagus Nerve.
- Innervates the Intrinsic Muscles of the Larynx.

Spinal Nerve Root

- Emerges from the Ventral Horn of the Upper Cervical Spinal Cord.
- Innervates the STERNOCLEIDOMASTOID (SCM) and UPPER TRAPEZIUS Muscles.

Nuclei Location

Cranial Nerve Root

- Medulla.

Spinal Nerve Root

- C1 – C5 Spinal Cord Levels in the Ventral Horn.

Function

Cranial Nerve Root

- Controls Elevation of the Larynx During Swallowing.

Spinal Nerve Root

- Innervation of the SCM Muscle allows for:
  - HEAD ROTATION to CONTRALATERAL SIDE
  - HEAD FLEXION/EXTENSION
- Innervation of the UPPER TRAPEZIUS M allows for:
  - SHOULDER ELEVATION and SHOULDER FLEXION ABOVE 90 degrees
  - The ACCESSORY and VAGUS Ns are the ONLY CNs that Innervate Organs and Glands BELOW the HEAD

Lesion Symptoms

Cranial Nerve Root Lesions

- Dysphagia secondary to decreased Laryngeal Elevation.

Spinal Nerve Root Lesions

- Weakness Rotating the Head to the Contralateral Side (because the Ipsilateral SCM Muscle that is affected functions to rotate the head to the opposite side).
- Weakness Flexing the Head Laterally and Forward (SCM).
- Weakness Extending the Head (SCM).
- Weakness Elevating the Shoulder (shrugging the shoulder) on the Ipsilateral side (Upper Trapezius Muscle).
- Weakness Flexing the Arm above 90º (Upper Trapezius Muscle) on the Ipsilateral side.

Test

To Test Cranial Nerve Root

- Place Index and Middle Fingers over patient's Adam's Apple (Laryngeal Muscles).
- Ask patient to swallow.
- Check for Normal Rise and Fall of Larynx.
- Check for Presence of Dysphagia using procedures described in Glossopharyngeal N section.

To Test Spinal Nerve Root

- Test SCM Muscle as Follows:
  - Ask patient to Flex Head Laterally and Forward, and Rotate Head to the Opposite Side (on both the involved and uninvolved sides). Therapist instructs the patient to resist the therapist's attempts to prevent the desired movement.

- Compare both sides of the body to observe symmetry of movement and strength.
- Check for atrophy in involved SCM Muscles.
- Test UPPER TRAPEZIUS Muscles as Follows:
  - Ask patient to Shrug Both Shoulders toward Ears. Therapist instructs the patient to resist attempts to depress the elevated shoulders.
  - Check for symmetry in shoulder movement and strength.
  - Check for atrophy of involved Upper Trapezius Muscles.

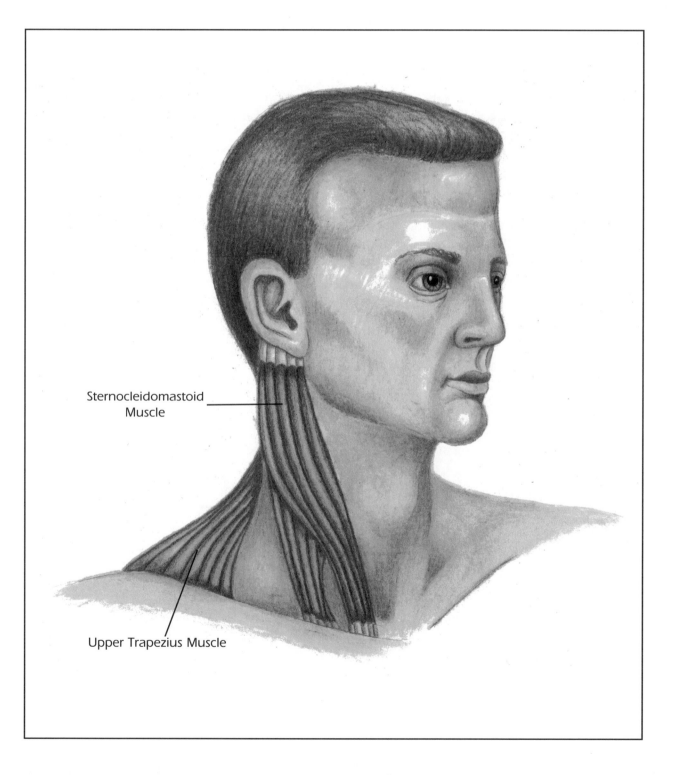

Sternocleidomastoid Muscle

Upper Trapezius Muscle

CN 12: Hypoglossal Nerve

Carries
- MOTOR Information.

Nuclei Location
- Medulla.

Function
- Innervates the Muscles of the Tongue. The Hypoglossal CN is responsible for TONGUE MOVEMENT.

Lesion Symptoms
- DYSARTHRIA secondary to impaired tongue musculature—inability to produce required movements for sound and word formation.
- IPSILATERAL DEVIATION of the TONGUE.
- DYSPHAGIA (because the Tongue Muscles are needed to manipulate food into a bolus in the mouth and propel bolus to the pharynx).
- IPISILATERAL ATROPHY of TONGUE. IPSILATERAL PARALYSIS of TONGUE.

Test
- Ask Patient to Protrude Tongue.
- Note whether Tongue deviates to the Lesion Side.
- Note whether there is Unilateral or Bilateral Atrophy of the Tongue Muscles.
- Check for Tongue Tremors or Involuntary Tongue Movements.
- Ask Patient to Move Tongue from Side to Side.
- Check for asymmetry in movement and weakness.
- Ask Patient to Push Tongue Against Both Cheeks.
- Ask patient to resist the therapist's attempts to depress the cheek as patient pushes the cheek outward.
- Note whether there is asymmetry in Tongue Strength.

THE STAGES OF SWALLOWING:
CRANIAL NERVES THAT MEDIATE THE FUNCTION OF SWALLOWING

There Are Three Stages of Swallowing
- ORAL
- PHARYNGEAL/LARYNGEAL
- ESOPHAGEAL

Cranial Nerves Involved in Swallowing
- CN 5   TRIGEMINAL N
- CN 7   FACIAL N
- CN 9   GLOSSOPHARYNGEAL N
- CN 10   VAGUS N
- CN 11   ACCESSORY N
- CN 12   HYPOGLOSSAL N

Stages of Swallowing

Stage	Function	CN
ORAL	1. Food is brought into the mouth; the lips close.	1. CN 7
	2. Jaw, cheek and tongue movements manipulate food into a bolus.	2. CN 5, 7, 12
	3. Tongue moves bolus to pharynx.	3. CN 12
	4. Larynx closes.	4. CN 10
	5. Swallow reflex is triggered.	5. CN 9
PHARYNGEAL LARYNGEAL	1. Food moves into pharynx.	1. CN 9
	2. Soft palate rises to block food from entering the nasal cavity.	2. CN 10
	3. Epiglottis covers trachea to prevent food from entering the lungs.	3. CN 10
	4. Pharynx rises and falls during swallowing.	4. CN 11
	5. Peristalsis moves food to the esophageal entrance, the sphincter opens and food moves into the esophagus.	5. CN 10
ESOPHAGEAL	Peristalsis moves food into the stomach.	CN 10

Sensory Receptors

- A SENSORY RECEPTOR is a specialized nerve cell that is designed to respond to a specific Sensory Stimulus (eg, Touch, Pressure, Pain, Temperature, Light, Sound, Position in Space).
- Sensory Receptors only accept molecules that have a complimentary receptor site organization.

THREE TYPES OF SENSORY RECEPTORS

Exteroceptors
- A Sensory Receptor that is adapted for the reception of stimuli from the EXTERNAL WORLD (outside of the body).
- Example: Visual, Auditory, Tactile, Olfactory, and Gustatory Receptors.

Interoceptors
- Receive sensory information from INSIDE the body—eg, from the viscera (hollow organs and glands).
- Interoceptors detect internal bodily sensations—such as stomach pain, pinched spinal nerves, or inflammatory processes in the deep layers of the skin.

Proprioceptors
- Proprioceptors are sensory receptors located in the Muscles, Tendons, and Joints of the body, and in the Utricles, Saccules, and Semicircular Canals of the inner ear (ie, the Labyrinths of the inner ear).
- These detect body position and movement.

DEVELOPMENTAL CLASSIFICATION OF SENSORY RECEPTORS

Protopathic
- Considered to be old phylogenetically.
- Poorly Localized. Protopathic Receptors are adapted to identify gross bodily sensation, rather than precise regions of sensation.
- Protopathic Receptors detect CRUDE TOUCH and DULL PAIN rather than Discriminative Touch and Sharp Pain.
- Evolutionary Function: enables the organism to detect possible (but not imminent) danger in the environment.

Epicritic
- Considered to be newer phylogenetically.
- Epicritic Receptors can detect ACUTE SENSIBILITY—such as DISCRIMINATIVE TOUCH, SHARP PAIN, EXACT JOINT POSITION, and the EXACT LOCALIZATION OF A STIMULUS.
- Evolutionary Function: allows the organism to explore the environment in precise detail.

Children with Sensory Processing Problems

- In children with sensory processing problems, the Epicritic and Protopathic Sensory Systems may be dysfunctional.

A Child with a Dominant Protopathic System

- May not be able to receive adequate sensory stimulation from the environment; or the sensory data that enter is dulled. It would be as if the child received all sensory data filtered through gloves, earmuffs, and a heavy winter coat that decreased the child's direct sensory stimulation from the environment.
- Some children with autism may have a Dominant Protopathic System. They may seek out a great deal of sensory stimulation in an attempt to compensate for their inability to receive adequate sensory stimulation from the environment.

A Child with a Dominant Epicritic Sensory System

- Such children experience heightened sensation from the environment. Certain sensations may be experienced as painful or intolerable. This is called TACTILE DEFENSIVENESS. For example, a child may experience the feel of water on his or her body as painful and refuse to take showers. Some children refuse to wear anything but shorts in the winter because they cannot tolerate the feel of clothing on their body.

Sensory Receptors Classified by Anatomical Location

Cutaneous Receptors

- Respond to Pain, Temperature, Pressure, Vibration, and Discriminative Touch.
- Found in the Superficial and Deep Layers of Skin.
- Cutaneous Receptors are also classified as Exteroceptors (coming from an external source) or Interoceptors (eg, Inflammatory Process in the Deep Layers of the Skin).

Muscle, Tendon, and Joint Receptors

- Located in the Muscles, Tendons, and Joints of the body.
- Detect Muscle Length, Muscle Tension, Joint Position, Deep Muscular and Joint Pain, and Tendonitis.
- These sensory receptors are also classified as PROPRIOCEPTORS.

Visceral Receptors

- Respond to Pressure and Pain from the Internal Organs.
- Also considered to be Interoceptors.

Special Sense Receptors

- Visual Receptors: Rods and Cones of the Retina. Considered an Exteroceptor.
- Olfactory Receptors: Hair Cells located in the Mucous Lining of the Nasal Canal. Considered an Exteroceptor.
- Auditory Receptors: Hair Cells of the Cochlea. Considered an Exteroceptor.
- Gustatory Receptors: Taste Buds on the Tongue. Considered an Exteroceptor.
- Equilibrium: Semicircular Canals, Utricles, and Saccules of the Inner Ear. Considered Proprioceptors.

Sensory Receptors Classified by Structural Design

Mechanoceptors

- Stimulated by Mechanical Deformity.
- Example: Hair Cells of the Labyrinth System, Some Receptors of the Skin, and Stretch Receptors of the Skeletal Muscles.
- Detect Touch, Pressure, Vibration, Proprioception, Equilibrium, and Audition.

Thermal Receptors

- Detect changes in Temperature.
- Located all over an organism's external and internal body.

Chemoreceptors

- Respond to the presence of a particular Chemical.
- Example: Chemoreceptors are the sensory receptors involved in Olfaction and Gustation.

Photoreceptors

- Detect Light on the Retina of the Eye.

Receptor Fields

- A Receptor Field is a Body Area that Contains Specific Types of Sensory Receptor Cells.
- When stimulated, the Receptors on a Specific Receptor Field become Activated.

Location of Small and Large Receptor Fields

Small Receptor Fields

- Located on Body Areas with the Greatest Sensitivity.
- Lips, Hands, Face, Soles of Feet.

Large Receptor Fields
- Located on Legs, Abdomen, Arms, Back.

Function of Small vs Large Receptor Fields

Small Receptor Fields
- Fine Discrimination.
- Necessary for the Exploration of the Environment.

Large Receptor Fields
- Gross Discrimination.

CORTICAL REPRESENTATION OF RECEPTOR FIELDS (THE HOMUNCULUS)

- Small Receptor Fields (eg, those of the Mouth, Tongue, Hands, Feet) are given a Large Amount of Cortical Representation on both the Sensory and Motor Homunculi.
- Large Receptor Fields (ie, Legs, Abdomen, Arms, Back) are given Smaller Amounts of Cortical Representation on the Homunculi.

Neurons and Action Potentials

NEURONS AND ACTION POTENTIALS

Neuron
- A Neuron is the Electrically Excitable Nerve Cell and Fiber of the Nervous System.
- Composed of:
  - A CELL BODY (soma) with a NUCLEUS, DENDRITES, A MAIN AXON BRANCH (with possible axon collaterals) and TERMINAL BOUTONS.

Cell Body
- Contains the Nucleus of the Neuron, which stores the genetic codes of the organism.

Dendrites
- The Tree-Like Processes that attach to the Cell Body.
- Receive messages from the Terminal Boutons of a presynaptic neuron.
- Dendrites can Bifurcate, or produce additional Dendritic Branches.
- Bifurcation increases the Neuron's Receptor Sites.

Axon Hillock
- The Region where the Cell Body and Axon Attach.

Axon
- Fiber emerging from the Axon Hillock and extending to the Terminal Boutons.
- Axons transmit Action Potentials, or nerve signals, to the Terminal Boutons.

Myelin
- Axons are covered by a Cellular Sheath called Myelin.
- Composed of Lipids and Proteins.
- Acts as an insulating substance of the Axon and serves to conduct nerve signals.
- The more Myelin the Axon has, the faster its Conduction Rate.
- Two Primary Types of Myelin:
  - SCHWANN CELLS
    - In peripheral nervous system (PNS).
    - Able to Regenerate if damaged.
  - OLIGODENDRITES
    - In central nervous system (CNS).
    - Cannot Regenerate if damaged.

Nodes of Ranvier
- Spaces between the Myelin where nerve signals jump from one node to the next in the process of conduction.

Axon Collaterals
- Project from the Main Axon Structure.
- Serve to transmit nerve signals to several parts of the nervous system simultaneously.

Terminal Boutons (Synaptic Boutons)
- Emerge from the End branches of the Axon and contain the Neurotransmitter Substances.

Synaptic Cleft
- Space between a Presynaptic Neuron's Terminal Boutons and a Postsynaptic Neuron's Dendrites.
- The Terminal Boutons release their Neurotransmitter substances into the Synaptic Cleft.

Neurotransmitter
- Chemical stored in the Terminal Boutons.
- Released into the Synaptic Cleft to transmit messages to another neuron.

Presynaptic Neuron
- First Order Neuron that releases its Neurotransmitter into the Synaptic Cleft.

Postsynaptic Neuron
- Second Order Neuron.
- Receives the Presynaptic Neuron's Neurotransmitter Substance from the Synaptic Cleft, but only if the Neurotransmitter possesses the Specific Molecules that can Bind to the Postsynaptic Neuron's Receptor Sites.

Re-Uptake Process
- Process by which a Neurotransmitter is reabsorbed into the Presynaptic Neuron's Terminal Boutons.

Down Regulation
- Process by which a Neurotransmitter is absorbed into a Postsynaptic Neuron's Dendrites and Cell Body.

Synaptic Delay
- The time required for the Neurotransmitter to Diffuse across a Postsynaptic Neuron's Membrane.
- Average time is 1 to 2 milliseconds.

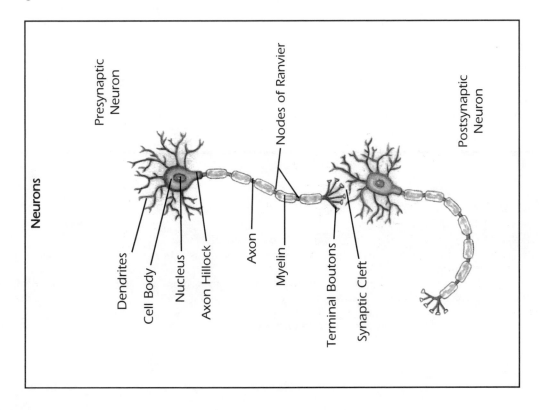

Antitransmitter

- Chemical substance that breaks down a Neurotransmitter so that the Postsynaptic Neuron can Repolarize in order to fire again.
- Antitransmitters terminate the Postsynaptic Neuron's Response.
- Located in the Synaptic Cleft.
- Example: ACETYLCHOLINESTERASE (AChE) is the Antitransmitter for Acetylcholine (ACh). ACh is a neurotransmitter important for activating Neuromuscular-Joint Movements.

Post-Tetanic Potentiation

- Occurs in synapses that are frequently used.
- When the Presynaptic Bouton becomes excited, it releases greater amounts of Neurotransmitter substance.
- The Postsynaptic Neuron then has Prolonged and Repetitive Discharge after firing due to too much Neurotransmitter Release or too slow Antitransmitter Work.

Decreases in Synaptic Transmission

Anoxia

- Lack of Oxygen.
- Synaptic Transmission begins to fail within 45 seconds of Anoxia.

Paralysis Due to Poison

- Paralysis due to poison—such as snake venom or curare atropine—occurs because the poison competes with ACh at the Neuromuscular Junction.
- This causes a decreased postsynaptic potential at the cell membrane of the muscle.
- Example: Botulism (food poisoning) is a condition in which the Release of ACh at the Presynaptic Bouton is blocked. This causes fatigue of the Neurotransmitters at the Neuromuscular Junction, resulting in Skeletal Paralysis.

Spasms Due to Cholinergic Drugs

- Cholinergic Drugs are chemicals that bind with and activate the same receptors as ACh.
- These drugs increase the effects of ACh at the Neuromuscular Junction, thus producing Spasms in the muscle.
- TARDIVE DYSKINESIA is a condition caused by Cholinergic Drugs. Patients may experience full body spasms, lip smacking, and tongue protrusion.

Synaptic Fatigue

- Occurs as a result of a Neurotransmitter Depletion due to the Repetitive Stimulation of a Presynaptic Neuron.
- Synaptic Fatigue underlies the Addictive Process. Every time an individual uses an addictive substance, the CNS is abnormally flooded with the Neurotransmitter Dopamine. This causes the brain to decrease its own natural production of Dopamine, leading to cravings for the addictive substance.

Multiple Sclerosis

- Characterized by Random Demyelination of the CNS.
- Signs and Symptoms:
  - Variable, as Demyelinated Lesions can occur in a wide variety of locations in the CNS.
  - Sensory Symptoms: Numbness, Paresthesias, Lhermitte's sign (causalgia radiating down the back or limbs—elicited by neck flexion).
  - Motor Symptoms: MS of the spinal cord produces asymmetrical weakness secondary to plaques that interfere with the descending motor tracts; Ataxia of the limbs secondary to an interruption of conduction in the Dorsal Columns.

Electrical Potentials at Synapses

- The Release of Neurotransmitters into the synaptic cleft results in the stimulation of receptors on the Postsynaptic Neuron's Cell Membrane.
- The Chemical Stimulation of these Receptors can cause the Membrane Ion Channel to Open.
- The flux of Ions across the Postsynaptic Membrane generates a Postsynaptic Potential.

Postsynaptic Potentials

- Postsynaptic Potentials are changes in Ion Concentration on the Postsynaptic Membrane.
- When a Neurotransmitter binds to a receptor site on a Postsynaptic Membrane, the result may be either DEPOLARIZATION or HYPERPOLARIZATION of the Postsynaptic Membrane.
- DEPOLARIZATION of the Cell Membrane causes the Neuron to become EXCITED. An EXCITATORY POSTSYNAPTIC POTENTIAL (EPSP) is generated.
- HYPERPOLARIZATION of the Cell Membrane causes INHIBITION of the Neuron. An INHIBITORY POSTSYNAPTIC POTENTIAL (IPSP) is generated.

Membrane Potentials

- The Membrane Potential is the Electrical Charge that travels across the Cell Membrane.
- It is the Difference between the Chemical Composition Inside and Outside the Cell—in other words, the Sodium Potassium Balance Inside and Outside the Cell.
- If the Membrane Potential is strong enough, it causes an Action Potential.

Action Potentials

- An Action Potential is the Brief Electrical Impulse that provides the basis for Conduction of Nerve Signals along the Axon.
- It results from the brief changes in the Cell's Membrane Permeability to Sodium and Potassium Ions.
- A strong enough Action Potential will cause the Neuron to become Excited and start the Conduction Process.

Sequence of Events of Synaptic Transmission

1. An Action Potential arrives at the Presynaptic Terminal Boutons.
2. The Membrane of the Presynaptic Terminal Boutons DEPOLARIZES, thus causing the Opening of the Voltage-Gated Ca++ Channels.
3. The Influx of Ca++ (in the Terminal Boutons) triggers the release of Neurotransmitters into the Synaptic Cleft.
4. If the Neurotransmitter has compatible molecules, it will bind to the Receptor Sites of the Postsynaptic Neuron.

Events of Excitatory Postsynaptic Potentials

1. Na+ Channels remain closed during the state of a Resting Membrane.
2. The Na+ Channels open when a Neurotransmitter, released into the synaptic cleft, binds to the Membrane Receptors of the Postsynaptic Cell Membrane.
3. The Resulting Influx of Na+ DEPOLARIZES the Membrane and causes Excitation of the Postsynaptic Neuron.

Events of Inhibitory Postsynaptic Potentials

1. Cl- Channels remain closed during the state of a Resting Membrane.
2. Cl- Channels open when a Neurotransmitter is released into the Synaptic Cleft and bind to Receptor Sites of the Postsynaptic Cell Membrane.
3. The Resulting Influx of Cl- HYPERPOLARIZES the Cell Membrane, thus causing Inhibition of the Postsynaptic Neuron.

Special Sense Receptors

- The SPECIAL SENSE RECEPTORS are designed to transmit sensory information for Olfaction, Gustation, Vision, Audition, and Equilibrium.

OLFACTION

Olfactory Receptors
- Cilia, or Hair Cells, located in the Nostrils and Nasal Membranes.

Physiology
- Some Chemical Stimulus (an odor) is received by the Hair Cell Receptors.
- The Odor Molecules dissolve in the mucous that bathes the receptors on the cilia.
- The dissolution of the molecules causes a Membrane Potential in the receptor endings.
- If the Membrane Potential is strong enough, an Action Potential will be generated.
- The Action Potential is then propagated down Cranial Nerve 1: The Olfactory Nerve.

Olfactory Pathway
- The Olfactory Pathway leads from the Nasal Membrane to the Olfactory Bulb and Tract, and travels to the Olfactory Cortex (Pyriform Cortex).
- The Olfactory Cortex is part of the Limbic System, along with the Amygdala and Hippocampus.
- The Amygdala and Hippocampus play roles in the storage of long-term memories.
- The Olfactory Cortex takes information from the Hippocampus and projects it to the Hypothalamus. From the Hypothalamus, the olfactory information is then projected to the Dorsomedial Thalamus, and finally to the Orbitofrontal Cortex for Conscious Association of the odor with previously stored memories.
- The connection between the Olfactory Pathway and the Hippocampus is phylogenetically old. This is the connection that triggers the remembrance of old memories in response to odors.
- The memories that are elicited as a result of the connection between the Olfactory Tract and Hippocampus have a strong emotional component. This is because the Limbic System is the Emotional Center of the Brain.

Therapeutic Significance
- Olfactory Stimulation is used with comatose patients to facilitate CNS arousal and activity.
- Aroma Therapy is based on the principle that odors can elicit enhanced moods as a result of the Limbic-Olfactory Connection.

Olfactory Lesions

A Bilateral Lesion of the Olfactory Nerve
- Will result in the loss of smell. Occurs frequently in patients with traumatic brain injury, secondary to Brainstem involvement.

A Lesion of the Olfactory Cortex (Pyriform Cortex)
- Results in the loss of smell.

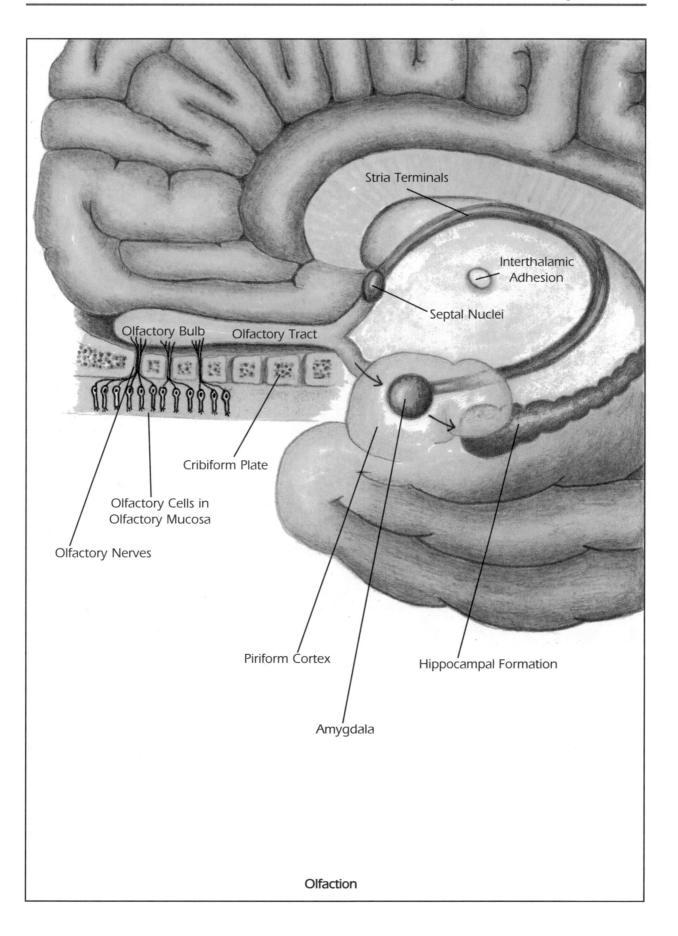

Stria Terminals

Interthalamic Adhesion

Septal Nuclei

Olfactory Bulb

Olfactory Tract

Cribiform Plate

Olfactory Cells in Olfactory Mucosa

Olfactory Nerves

Piriform Cortex

Hippocampal Formation

Amygdala

Olfaction

GUSTATION

Gustatory Receptors

- The Gustatory Receptors are the Tastebuds found on the folds of Papillae on the surface of the tongue.
- Papillae are the small protuberances on which the tastebuds lie.
- The tastebuds form synapses with the sensory neurons that convey taste information to the brain.

Physiology

- Saliva dissolves food in the mouth.
- Dissolved ions enter the pores of the papillae and tastebuds.
- This causes a Membrane Potential.
- If the Membrane Potential is strong enough it will cause an Action Potential.

Gustatory Pathway

- The Gustatory Pathway leads from the Tastebud Receptors to the Solitary Nucleus in the Medulla.
- Gustatory information is then propagated to the Amygdala, Hypothalamus, and Thalamus.
- From the Thalamus, Gustatory information then travels to the Primary Gustatory Cortex (located in the Frontal Insula) for conscious interpretation.

The Connection Between Taste and Smell

- The sense of smell is critical for the function of taste.
- If the sense of smell is lost, the sense of taste will also be lost.
- The Orbitofrontal Cortex seems to be a brain region that integrates Gustatory and Olfactory information.
- When the Orbitofrontal Cortex cannot integrate Olfactory and Gustatory information, Gustation is lost.
- Damage to the Orbitofrontal Region is common in TBI and may account for the lost sense of smell and taste that many individuals with TBI experience.

Therapeutic Significance

- Gustatory Stimulation is used in the sensory stimulation treatment of comatose patients.
- Also used to facilitate Oral Motor Function in children and adults with oral musculature dysfunction.

Gustatory Information Is Recieved From CNs 7, 9, and 10

- The taste receptors on the Anterior of the Tongue are mediated by CN 7.
- The taste receptors on the Posterior of the Tongue are mediated by CN 9.
- The taste receptors on the Palate and Epiglottis are mediated by CN 10.

Specific Taste Regions of the Tongue

- The Tip of the Tongue can detect all tastes, but predominantly detects Sweet tastes.
- Salty tastes are detected on the sides of the tongue.
- Bitter and Sour tastes are detected on the Posterior of the Tongue

Gustation

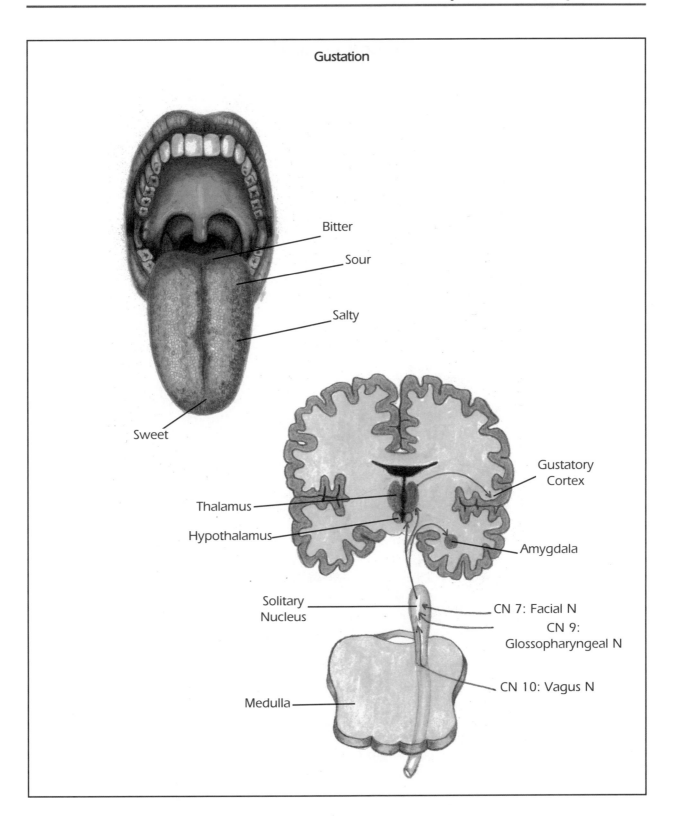

Bitter

Sour

Salty

Sweet

Gustatory Cortex

Thalamus

Hypothalamus

Amygdala

Solitary Nucleus

CN 7: Facial N

CN 9: Glossopharyngeal N

CN 10: Vagus N

Medulla

Vision (Acuity)

Anatomy of the Eyeball

Sclera

- Outer layer of the eyeball.

Cornea

- The Transparent Anterior Portion of the Sclera is called the Cornea.
- The Cornea lies in front of the Iris and Pupil.

Choroid

- The Choroid is the Middle Layer of the eyeball.
- Contains the Iris and Lens.
- Extends behind the Retina.
    - IRIS
        - The Iris is the Circular Structure that forms the Colored Portion of the eye.
        - Controls the size of the pupil opening.
    - LENS
        - The Lens is the structure that focuses Light Rays on the Retina.

Anterior Chamber

- The Anterior Chamber contains the gelatinous fluid between the Cornea and Lens called Aqueous Humor.

Posterior Chamber

- The Posterior Chamber is located directly behind the Lens and extends to the retina.
- Contains both Aqueous Humor and Vitreous Humor.

Pupil

- The Pupil is the circular black opening in the center of the Iris.
- Light enters through the Pupil.
- The amount of light permitted is controlled by the constriction of the Iris.

Fovea Centralis

- The Fovea is the central region of the Retina.
- Primarily contains Cone Receptors.
- The Fovea is the location upon which incoming visual images are focused.

Retina

- The Retina is the Photosensitive Layer at the posterior of the eyeball.
- Contains the Rods and Cones—the Photoreceptors.
    - RODS
        - Specialized Receptors for Peripheral Vision.
        - Function optimally in Dim Light.
    - CONES
        - Specialized Receptors for Color Vision and Acuity.
        - Function optimally in Bright Light.
        - In Dim Light humans are color blind and lack acuity.

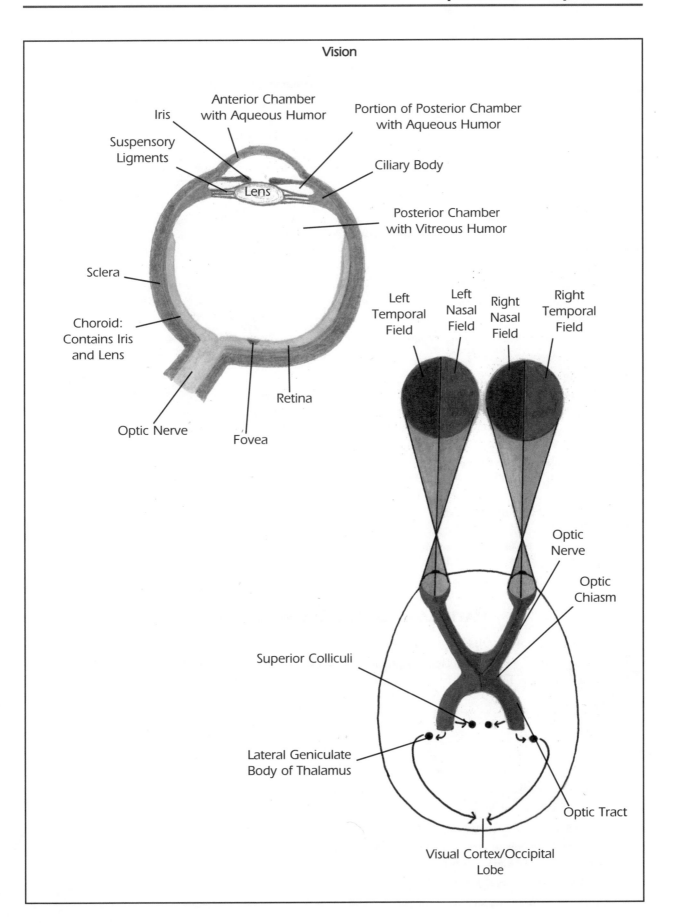

Vision

Iris

Anterior Chamber
with Aqueous Humor

Portion of Posterior Chamber
with Aqueous Humor

Suspensory
Ligments

Ciliary Body

Lens

Posterior Chamber
with Vitreous Humor

Sclera

Left
Temporal
Field

Left
Nasal
Field

Right
Nasal
Field

Right
Temporal
Field

Choroid:
Contains Iris
and Lens

Retina

Optic Nerve

Fovea

Optic
Nerve

Optic
Chiasm

Superior Colliculi

Lateral Geniculate
Body of Thalamus

Optic Tract

Visual Cortex/Occipital
Lobe

Visual Receptor Pathway

- Light Waves enter the eye and are refracted by the Cornea and Lens.
- The Light Waves are then focused upon either the Fovea (where the Cones are located) or the Peripheral Retina (where the Rods are located).
- A Chemical Reaction occurs in the Rods and Cones and a Membrane Potential is generated.
- If the Membrane Potential is strong enough, an Action Potential is then propagated along Cranial Nerve 2—the Optic Nerve.
- Visual Signals are then propagated down the Optic Nerve to the Optic Chiasm.
- The Optic Chiasm is where the Optic Ns join together at the base of the brain. The Optic Chiasm is a midline structure. The Chiasm joins to the Optic Tracts.
- Visual Signals then project from the Optic Tracts to 2 Locations: Superior Colliculi of the Midbrain and the Lateral Geniculate Nucleus of the Thalamus.
- The fibers that project to the Superior Colliculi then travel along the Medial Longitudinal Fasciculus, back to the Thalamus.
- All visual fibers that travel through the Thalamus then continue on to the Occipital Lobe.
- In the Occipital Lobe, visual signals first travel to the Primary Visual Cortex (V1). V1 is responsible for the Detection of a visual stimulus.
- Visual Signals then travel from V1 to the Visual Association Cortices for Interpretation of the visual stimulus.

Visual Field Pathways

- Visual Information from the Nasal Fields is Refracted by the Lens. As a result, visual information from the Nasal Fields travel along the lateral aspect of the Optic Nerves and Tracts.
- Visual Information from the Temporal Fields is also Refracted by the Lens. As a result, visual information from the Temporal Fields travel along the medial aspect of the Optic Nerves and Tracts.
- Visual Information from the Temporal Fields cross at the Optic Chiasm.
- The Lens also inverses the visual image on the retina. As a result, visual information from the Right Visual Field is focused on the Left Region of the Retina.
- Visual information from the Left Visual Field is focused on the Right Region of the Retina.
- Each Cerebral Hemisphere receives visual information from the Contralateral Visual Field.
- The Right Cerebral Hemisphere receives visual information from the Left Visual Field.
- The Left Cerebral Hemisphere receives visual information from the Right Visual Field.
- Thus, if the Left Visual Field is lost, it is due to a lesion in the Right Optic Tract or Right Cerebral Hemisphere. If the Right Visual Field is lost, it is due to a lesion in the Left Optic Tract or Left Cerebral Hemisphere.

PATHOLOGY OF THE VISUAL PATHWAYS

Contralateral Homonymous Hemianopsia

- A Contralateral Homonymous Hemianopsia is a loss of the visual field on the opposite side of the lesion.
- A Left Visual Field Cut, or Left Contralateral Homonymous Hemianopsia, results from a Lesion in the Right Optic Tract.
- A Right Visual Field Cut, or Right Contralateral Homonymous Hemianopsia, results from a lesion in the Left Optic Tract.

Bitemporal Hemianopsia

- A Bitemporal Hemianopsia occurs when the Temporal Fields in Both Eyes have been Lost.
- Results from a Lesion to the Central Optic Chiasm.
- Results in Tunnel Vision.

Blindness as a Result of Optic Pathway Damage

- Unilateral Blindness can result from a lesion to the Ipsilateral Optic Nerve.
- Bilateral Blindness can result from a large lesion that knocks out the entire Optic Chiasm.

A Blind Spot in the Visual Field

- A Blind Spot inside a Visual Field suggests Retinal Damage (NOT Optic Nerve or Tract damage).

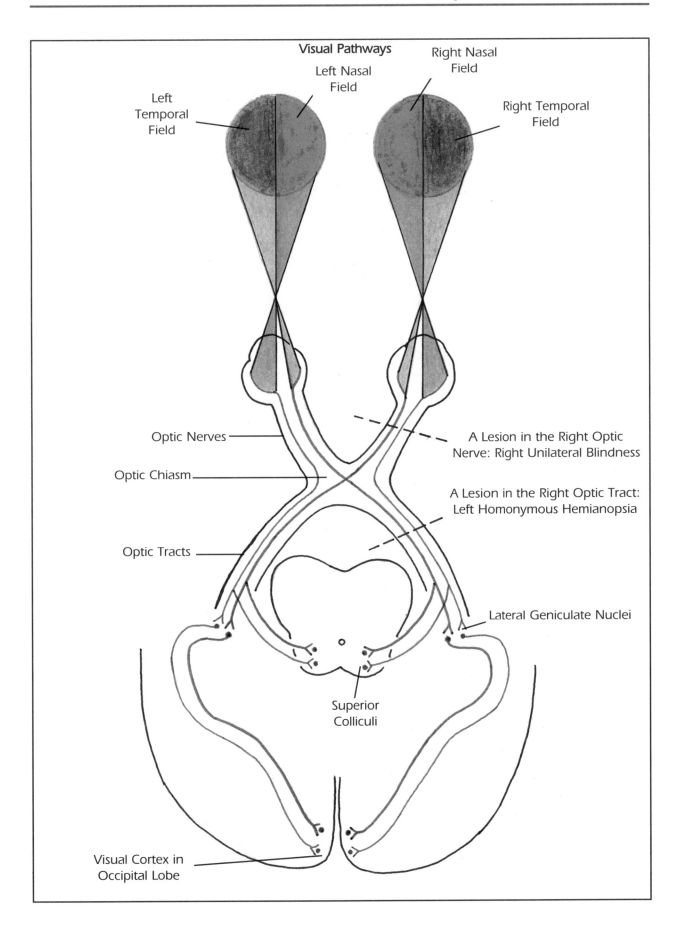

Visual Pathways

Left Nasal Field

Right Nasal Field

Left Temporal Field

Right Temporal Field

Optic Nerves

A Lesion in the Right Optic Nerve: Right Unilateral Blindness

Optic Chiasm

A Lesion in the Right Optic Tract: Left Homonymous Hemianopsia

Optic Tracts

Lateral Geniculate Nuclei

Superior Colliculi

Visual Cortex in Occipital Lobe

AUDITION

Auditory Receptor Anatomy

Cochlea

- The Cochlea is a fluid filled structure located in the Inner Ear.
- Contains a structure called the Organ of Corti.

Organ of Corti

- The Organ of Corti contains the Auditory Receptors—Hair Cells attached to the Basilar Membrane at the base of the Organ of Corti.

Auditory Pathway

- Sound, or Vibration, is the stimulus for the Organ of Corti.
- Sound waves travel through the External Auditory Meatus (the Ear Canal) to the Tympanic Membrane (the Eardrum).
- When sound waves reach the Tympanic Membrane, the membrane vibrates and causes the Ossicles to Vibrate.
- The Ossicles are the bones of the Middle Ear—the Malleus (Hammer), Incus (Anvil), and Stapes (Stirrup).
- The Stapes presses against the Oval Window causing it to vibrate.
- Cochlear vibration then causes the Basilar Membrane to flex back and forth.
- Pressure changes of the Cochlear Fluid are then transmitted to the Round Window.
- In response, the Round Window flexes back and forth in a manner opposite to the movement of the Oval Window.
- The Hair Cells of the Basilar Membrane begin to bend causing a Membrane Potential to be generated.
- If the Membrane Potential is strong enough, an Action Potential will be propagated down Cranial Nerve 8: the Vestibulocochlear Nerve.
- The Auditory Message is then sent to the Cochlear Nucleus in the Medulla. From here, fibers project to the Superior Olivary Nucleus (Medulla level), the Inferior Colliculus (Midbrain level), and to the Medial Geniculate Nucleus of the Thalamus.
- From the Thalamus, auditory fibers project to the Primary Auditory Cortex (A1) in the Temporal Lobe. A1 is responsible for the detection of auditory stimuli.
- The auditory messages are then sent to the Auditory Association Areas (A2+) for interpretation.

Hearing Impairment

- SENSORINEURAL: Involves damage to the Inner Ear, Vestibulocochlear Nerve, and/or the Brain.
- CONDUCTIVE: Involves damage to the Outer and/or Middle Ear Structures.

Lesion to the Auditory Portion of the Vestibulocochlear Nerve

- Causes Deafness or Tinnitus.

Lesion to the Primary Auditory Cortex

- Causes Cortical Deafness.

Lesion to the Auditory Association Cortices

- May cause Auditory Agnosia—inability to interpret sounds.

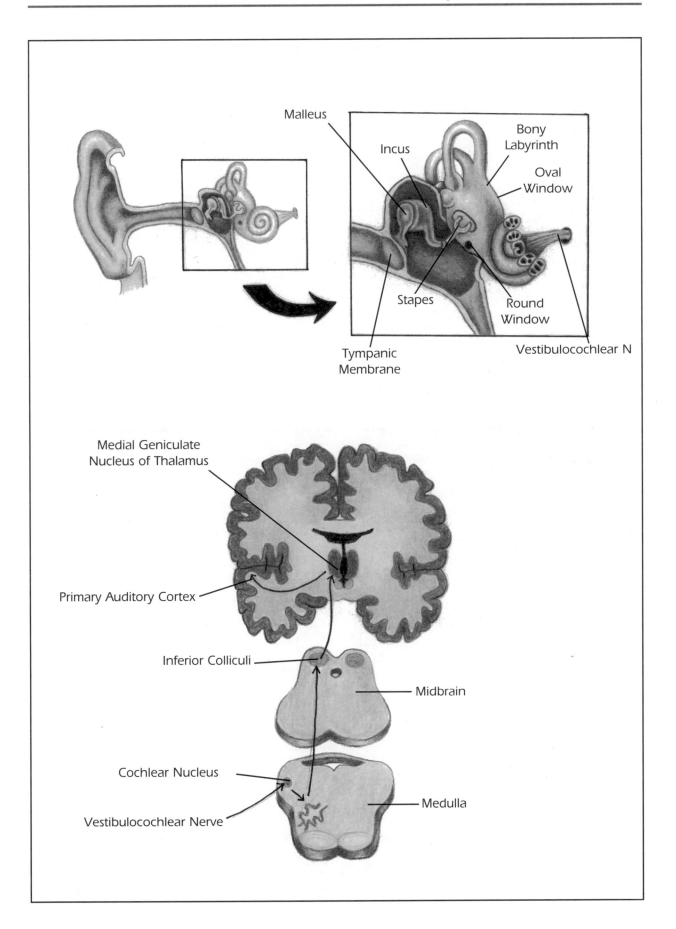

Malleus

Incus

Bony Labyrinth

Oval Window

Stapes

Round Window

Tympanic Membrane

Vestibulocochlear N

Medial Geniculate Nucleus of Thalamus

Primary Auditory Cortex

Inferior Colliculi

Midbrain

Cochlear Nucleus

Vestibulocochlear Nerve

Medulla

EQUILIBRIUM

Equilibrium Receptor Anatomy
- The Receptors of Equilibrium are the Hair Cells of the Semicircular Canals, Utricle, and Saccule located in the Inner ear.

Three Semicircular Canals
- The Three Semicircular Canals respond to Movement of the Head.
- Superior Semicircular Canal, Posterior Semicircular Canal, Horizontal Semicircular Canal.
- The Semicircular Canals are a system of canals called the Bony Labyrinth.
- Within the Bony Labyrinth is a Membranous Labyrinth.
- Both Labyrinths contain Fluid.
- Perilymph is the Fluid located within the Bony Labyrinth.
- Endolymph is the Fluid located within the Membranous Labyrinth.

Ampulla
- Located at the end of each Semicircular canal is a small enlargement called the Ampulla.
- Each Ampulla contains Hair Cells.

Utricle
- The Utricle is located just medially to the Semicircular Canals.

Saccule
- The Saccule is located just medial and inferior to the Utricle.

---

*Together, the **UTRICLE** and **SACCULE** respond to:*
- GRAVITY and CHANGES in HEAD POSITION.

*The **SEMICIRCULAR CANALS** respond to:*
- ANGULAR ACCELERATION AND DECELLERATION (Rotation).
- LINEAR ACCELERATION AND DECELLERATION (moving forward on a bike or in a car).

POST-ROTARY NYSTAGMUS

- Nystagmus is an involuntary, rapid, repetitive, jerky oscillating movement of the eyeballs in either a horizontal, vertical, or rotary direction.
- Nystagmus can occur in the absence of pathology in response to:
  - Rotation
  - Optokinetic Nystagmus (eg, being on a train and watching sequential telephone poles pass by)
  - Caloric testing
- Nystagmus can also occur as a result of Pathology:
  - Damage to the Labyrinths of the Inner ear
  - Vestibular Nuclei Damage at the Pons-Medulla Junction
  - Vestibulocochlear Nerve Damage
  - Extraocular Nuclei Damage in the Midbrain and Pons
  - Extraocular Cranial Nerve Damage
  - Cerebellar Damage
- Post-Rotary Nystagmus occurs normally after rotation. Patients with neurologic deficits will experience post-rotary nystagmus for longer periods after the rotation has stopped.
- The individual is rotated to the right (on a rotation board/swing).
- Fluid inside the labyrinths displaces towards the left—due to the sudden rotation of the head to the right.
- This causes the Hair Cells to Bend and become excited.
- An Action Potential is generated along the Vestibular Portion of the Vestibulocochlear Nerve.

- While the individual is rotating to the right, the fluid in the Labyrinths eventually catches up to the Labyrinths and begins to move at the same rate as the Labyrinths.
- This causes the Hair Cells to STOP firing—they are no longer being bent.
- The Rotation Board/Swing is stopped.
- The Fluid in the Labyrinths now displaces to the right—because the head has stopped moving but the fluid has not stopped yet.
- This again Bends the Hair Cells and causes them to become Excited.
- An Action Potential is generated and travels along the Vestibular Portion of the Vestibulocochlear Nerve.
- The Individual experiences Nystagmus to the Right and has a sensation of the room rotating to the right.
- If asked to point at a fixed object, the individual will overshoot it by pointing further right (past pointing to the right).
- The individual also experiences a tendency to fall to the right.

NYSTAGMUS TESTING

- There are several standardized tests to assess whether nystagmus is pathological or normal.

Post-Rotary Nystagmus Test (Section of the Sensory Integration and Praxis Test)
- This is a standardized test for children and adults.
- The individual sits on a rotating swing. The rotation is stopped and the therapist examines the length of the nystagmus.
- Eight to 14 seconds of nystagmus is a normal range for nystagmus to continue after rotation has stopped.

Caloric Testing
- Usually performed by physicians.
- Warm or cold water is squirted into the ear to induce nystagmus.
- May induce nausea due to the connection between the Vestibular System and the Vagus Nerve.

Optokinetic Testing
- Can be induced from prolonged, recurrent stimulation of the Extraocular Cranial nerves.
- Examples: Black and White Rotating Drum, Watching Telephone Poles pass by while seated on a train.

Electronystagmagraph
- Performed by physicians.
- Electrodes are placed on the Extraocular Eye Muscles.
- The individual is rotated; the rotation is stopped.
- The ENG records the nystagmus on paper (like an electromyograph).

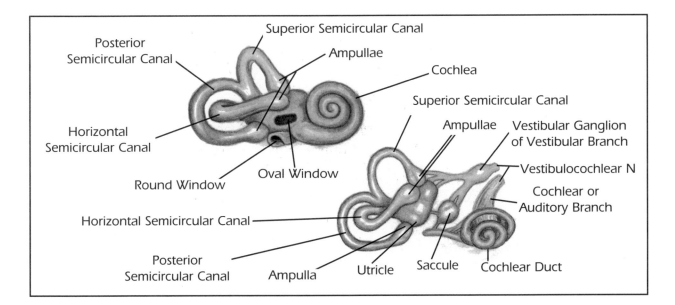

Vestibular System

FUNCTION

- Maintains Equilibrium and Balance.
- Maintains the Head in an Upright Vertical Position.
- Role in the Coordination of Head and Eye Movements.
- Influences Tone through the Alpha and Gamma Motor Neurons and the Medial and Lateral Vestibulospinal Tracts.

RECEIVES INPUT FROM

- The Vestibulocochlear Nerve.
- The Vestibular Nuclei in the Pons-Medulla Junction.
- The Vestibular Apparatus in the Inner Ear (Semicircular Canals, Utricle, and Saccule).
- The Cerebellum.
- The Extraocular Cranial Nerves and Nuclei.

VESTIBULAR PATHWAY

- The Vestibular Nuclei in the Pons-Medulla Junction receive Vestibular Information about Balance and Equilibrium from:
  - The Vestibular Apparatus in the Inner Ear
  - The Extraocular Nuclei in the Midbrain and Pons
  - The Cerebellum
- The Vestibular Nuclei send the information it receives to the Alpha and Gamma Motor Neurons in the Ventral Horn of the Spinal Cord.
  - This information is sent via the Vestibulospinal Tracts
- The Motor Neurons in the Ventral Horn synapse with Spinal Nerves in the Periphery.
- The Spinal Nerves in the Periphery project to Antigravity Muscles—the muscles that maintain the body's upright position against gravity. These are the Extensors in the Legs, Trunk, and Back.
- The Muscle Spindles, Golgi Tendon Organs, and Joint Receptors then send information about the body's position in space back to the Cerebellum in a Feedback Loop.
- The Cerebellum uses this information to make ongoing decisions about the modification of muscular activity, in order to enhance balance.
- There are also Motor Neurons in the Ventral Horn that project to the Head and Neck Musculature via the Medial Longitudinal Fasciculus.
  - These structures mediate the Head Righting and Tonic Neck Reflexes—reflexes that remain present throughout the lifespan to maintain the Head in an Upright Position
- Eye and Head Movements are also integrated by the Medial Longitudinal Fasciculus.

- This Tract allows Feedback about the Head's position to be continuously sent back to the Extraocular Nuclei in the Midbrain and Pons

RELATIONSHIP BETWEEN THE RETICULAR FORMATION AND THE VESTIBULAR SYSTEM

- The Reticular Formation is diffusely located in the Brainstem.
- Composed of the Reticular Activating and Inhibiting Systems.
- The Reticular Activating System screens sensory information to alert the Brain to attend to important incoming sensory data. Also responds to Excitatory Vestibular Stimulation by arousing the Brain and Body.
- The Reticular Inhibiting System acts as a tool to calm the Brain/Body in response to Inhibitory Sensory Information.
- The Reticular Formation in the Brainstem integrates information from the Vestibular System via the Medial Longitudinal Fasciculus Tract and the Vestibulospinal Tracts.
- Vestibular Sensory data—such as slow rocking—are integrated by the Reticular Formation and calm the individual.
- Excitatory Vestibular Sensory Data—such as fast dancing and roller coasters—are integrated by the Reticular Formation and arouse the Brain/Body.
- Children seek this kind of Inhibitory and Excitatory Vestibular Sensation to facilitate the development and organization of their nervous systems.
- The need for intense Vestibular Stimulation decreases as humans age. Often, adults cannot tolerate the same kind of intense Vestibular Stimulation they craved as children and adolescents.

RELATIONSHIP BETWEEN THE AUTONOMIC NERVOUS SYSTEM AND THE VESTIBULAR SYSTEM

- The ANS mediates Cranial Nerve 10, the Vagus Nerve.
- The Vagus Nerve has connections to the Vestibular Pathways.
- This connection accounts for why over-excitation of the Vestibular System can induce nausea and vomiting.

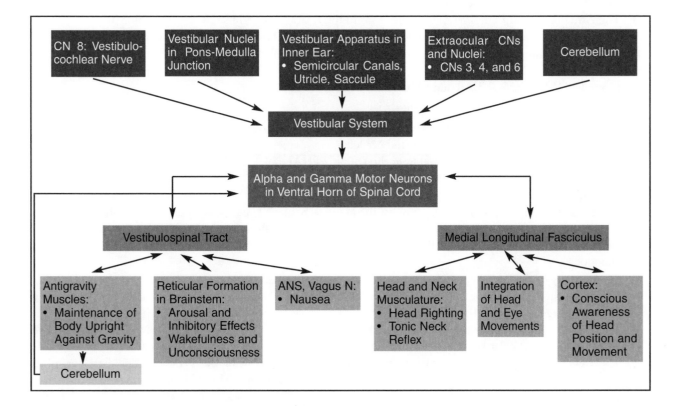

Autonomic
Nervous System

- The ANS innervates the INTERNAL ORGANS, BLOOD VESSELS, and GLANDS.
- Regulates CARDIAC and SMOOTH MUSCLE (muscle of glands and organs).
- Regulates SECRETION from GLANDS.
- Controls VEGETATIVE FUNCTIONS:
  - TEMPERATURE, DIGESTION, HEART RATE, and MAINTAINS HOMEOSTASIS OF INTERNAL ORGANS

CENTRAL COMPONENTS OF AUTONOMIC NERVOUS SYSTEM

- Central Portion of the ANS consists of parts of the CEREBRAL CORTEX, HYPOTHALAMUS, THALAMUS, LIMBIC SYSTEM, CEREBELLUM, and SPINAL CORD.
- Efferent fibers that originate in the Cortex project descending fibers through the Thalamus and Hypothalamus. These fibers end on a Cranial Nerve Nuclei to influence Involuntary Muscles, Vessels, and Glands.

Anterior and Posterior Hypothalamus

- The Anterior Hypothalamus sends off projections to the PARASYMPATHETIC Nervous System.
- The Posterior Hypothalamus sends off projections to the SYMPATHETIC Nervous System.
- The Hypothalamus exerts Autonomic Regulation over various BRAINSTEM CENTERS that control VEGETATIVE FUNCTIONS.
- Example: CADIOVASCULAR and RESPIRATORY FUNCTIONS.

Brainstem Centers: The Reticular Formation

- The Reticular Formation consists of interconnected neurons throughout the Brainstem.
- Located throughout the Midbrain, Pons, and Medulla.
- Sends and Receives projections to and from the DIENCEPHALON, CORTEX, and SPINAL CORD.
- Involved in the CONTROL OF POSTURE, VISCERAL MOTOR FUNCTION, SLEEP, AROUSAL/WAKEFULNESS.
- The Reticular Formation is an Ancient part of the Nervous System. The brains of primitive vertebrates are made up primarily of a reticular-type formation designed for survival functions.
- There are RESPIRATORY and CARDIOVASCULAR Centers in the Reticular Formation. These control Vital Functions.
- Severe injury to the Brainstem often results in death or a poor prognosis.
- If the Brainstem is left intact, but the Cortex is no longer active, one can still survive—considered to be in a Persistent Vegetative State (because the Brainstem controls vegetative functions).
- The Reticular Formation also receives sensory fibers from the Somatic and Visceral systems (including vision and olfaction). The Reticular Formation then sends this sensory information to the Thalamus and Cortex.
- The Reticular Formation's Motor Fibers synapse with Motor Neurons of the Pyramidal and Extrapyramidal Systems, and Motor Neurons that synapse with Preganglionic Autonomic Motor Neurons.
- The Reticular Formation is the origin of the Reticulospinal Tracts—descending extrapyramidal tracts of the Spinal Cord that regulate inhibition and facilitation of Antigravity Muscles.

Reticular Activating System

- The RAS is the portion of the Reticular Formation that is responsible for AROUSAL, ALERTNESS, and WAKEFULNESS.
- Filters all incoming sensory information. The RAS alerts the Cortex to attend to important Sensory Input.
- This results in a sharpening of ATTENTION to important sensory information from the environment.
- The RAS is diffusely located throughout the Brainstem, but is believed to be primarily located in the Midbrain.

Lesions to the Reticular Activating System

- Result in a Stuporous State. When the RAS is lost, the reticular inhibitory system (RIS) becomes dominant and causes heightened somnolence.

Reticular Inhibitory System

- The RIS is responsible for calming the organism. Certain types of sensory input—slow rocking, deep pressure—will activate the RIS to calm the body.
- It is believed that the RIS is primarily located in the Pons.

Lesions to the Reticular Inhibitory System

- Result in Constant Wakefulness. When the RIS is lost, the RAS becomes dominant and causes heightened arousal.

Limbic Lobe

- The Limbic Lobe has a role in the relationship between our emotions and the ANS.
- Heightened Emotional States will cause the Sympathetic Nervous System to become activated and dampen Parasympathetic Nervous System activity.
- When humans are upset or excited, the desire to eat is reduced—because the Sympathetic Nervous System is geared up and has shut down the Digestive System (regulated by the Parasympathetic Nervous System).
- When humans or animals are frightened, loss of bladder control can occur—fear can interfere with ANS regulation.
- When individuals are in great pain, nausea may occur as a result of the connections between the ANS, the Vagus Nerve, and the Limbic Lobe.

Cerebral Cortex

- The Cerebral Cortex interprets Sensory Information from the Reticular Formation and will modulate the Reticular Formation's effect on the body.

PERIPHERAL COMPONENTS OF AUTONOMIC NERVOUS SYSTEM

- The Peripheral Components of the ANS consist of PRE- AND POSTGANGLIONIC FIBERS that innervate the Viscera.

Preganglionic Fibers

- A Ganglion is a collection of Neurons in the peripheral nervous system (PNS).
- Preganglionic Fibers are Presynaptic Neurons (or primary neurons).
- Preganglionic Fibers have their Cell Bodies in the Brainstem (CNs 3, 7, 9, 10, 11).
- Preganglionic Fibers also have their cell Bodies in the Spinal Cord—in the Intermediolateral Horn of the Thoracic sections and first two Lumbar Sections.
- The Preganglionic Fibers extend from their Cell Bodies in the Brainstem or Spinal Cord and project to the periphery.
- In the periphery, the Preganglionic Fibers synapse on Postganglionic Neurons.

Postganglionic Fibers

- These are also called Postsynaptic or Second-Degree Neurons.
- The Postsynaptic Ganglionic Fibers are located in the periphery.
- Their Cell Bodies are located in the Autonomic Ganglia (groups of ANS organs and tissues).

Sympathetic Nervous System

- The Sympathetic Nervous System's Preganglionic Neurons have their Cell Bodies located in the Intermediolateral Horn of the Thoracic and first two Lumbar Sections of the Spinal Cord.

Parasympathetic Nervous System
- The Parasympathetic Nervous System's Preganglionic Fibers emerge from the Brainstem and Sacral Spinal Cord.

SYMPATHETIC NERVOUS SYSTEM

Function
- The Sympathetic Nervous System activates the FIGHT/FLIGHT RESPONSE, including INCREASED HEART RATE, BLOOD PRESSURE, ENERGY MOBILIZATION, and DECREASED DIGESTIVE AND REPRODUCTIVE FUNCTIONS.

Location of Cell Bodies
- The Sympathetic Nervous System is also called the THORACOLUMBAR Division because its Preganglionic Neurons have their Cell Bodies located in the Intermediolateral Horn of the Thoracic and first two Lumbar Sections of the Spinal Cord.

Sympathetic Chain Ganglia
- The Sympathetic Chain Ganglia are a series of interconnected sympathetic ganglia that lie adjacent to the vertebral column (on both sides).
- The Chain Ganglia receive input from the Preganglionic Sympathetic Fibers.
- The Chain Ganglia project Postganglionic Sympathetic Fibers to specific Target Organs and Tissues.

Preganglionic Sympathetic Fibers
- Extend from their Cell Bodies (in the Intermediolateral Horn of T1 – L2) to the Chain Ganglia.

Postganglionic Sympathetic Fibers
- Extend from the Chain Ganglia to (one of three) Collateral Ganglia (located outside of the chain).
- A Collateral Ganglion is a collection of Cell Bodies located outside of the Sympathetic Chain Ganglia.
- There are three main Collateral Ganglia in the Body (three on each side of the Vertebral Column):
  - CELIAC GANGLION
  - SUPERIOR MESENTERIC GANGLION
  - INFERIOR MESENTERIC GANGLION
- The Postganglionic Sympathetic Fibers ascend or descend through the Chain Ganglia before synapsing on one of the Collateral Ganglion.
- After synapsing on one of the Collateral Ganglion, the Postganglionic Sympathetic Fibers project to a Target or End Organ.

Neurotransmitters of the Sympathetic Division
- The Preganglionic Fibers use ACh.
- The Postganglionic Fibers use Noradrenalin and Norepinephrine.

PARASYMPATHETIC NERVOUS SYSTEM

Function
- Conservation and Restoration of Energy Stores.
- Maintains HEART RATE, RESPIRATION, METABOLISM, and DIGESTION in a state of Homeostasis.

Location of Cell Bodies
- Also called the CRANIOSACRAL Division because the Parasympathetic Nervous System's Cell Bodies are located in the Brainstem and Sacral Spinal Cord.

Preganglionic Parasympathetic Fibers
- The Preganglionic Parasympathetic Fibers extend directly from their Cell Bodies to the Terminal Ganglia located within or very near the organs that they supply.

Postganglionic Parasympathetic Fibers
- The Postganglionic Parasympathetic Fibers are short in length. They extend from the Terminal Ganglia to the Target End Organ.

Neurotransmitters of the Parasympathetic Division
- Both the Pre- and Postganglionic Fibers use ACh.

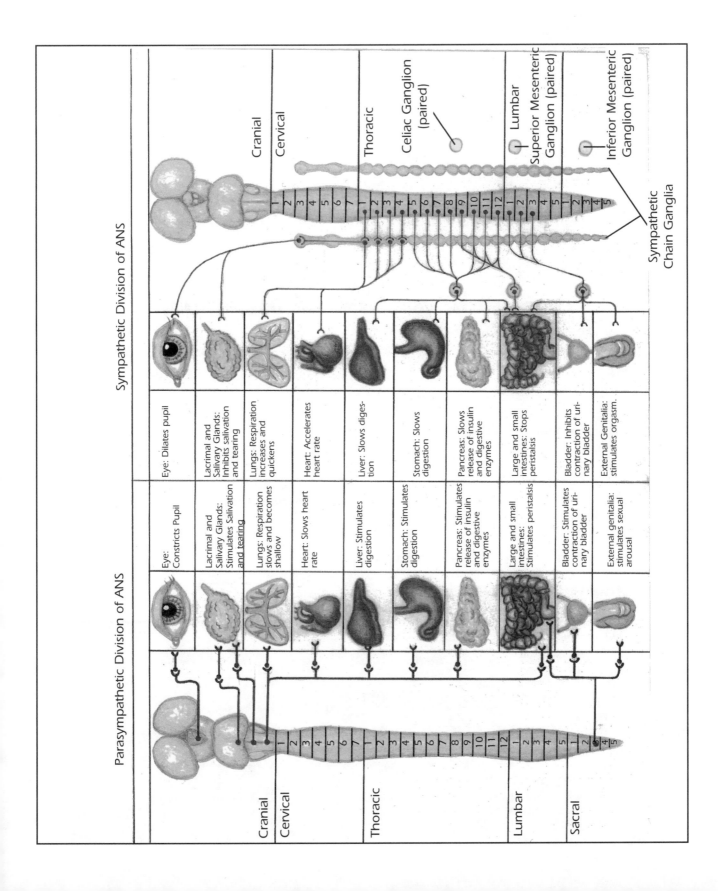

COMPARISON OF THE PARASYMPATHETIC AND SYMPATHETIC NERVOUS SYSTEMS

Function

- The Sympathetic Nervous System activates the Fight/Flight Response and gears the body up for action.
- The Parasympathetic Nervous System regulates Homeostasis and slows the body down.

Cell Bodies

- Sympathetic Nervous System Cell Bodies are located in the Intermediolateral Horn of the Thoracic and first two segments of the Lumbar Sections.
- Parasympathetic Nervous System Cell Bodies are located in the Brainstem and Sacral sections of the Spinal Cord.

Pathways

- Preganglionic Sympathetic Fibers extend from their Cell Bodies to the Sympathetic Chain Ganglia. Postganglionic Sympathetic Fibers then travel from the Chain Ganglia to a Collateral Ganglion, and finally synapse on a Target Organ.
- Preganglionic Parasympathetic Fibers extend from their Cell Bodies to a Terminal Ganglia. Postganglionic Parasympathetic Fibers then travel from the Terminal Ganglia to the Target Organ.

Neurotransmitters

- Preganglionic Sympathetic Fibers use ACh. Pre- and Postganglionic Parasympathetic Fibers use ACh.
- Postganglionic Sympathetic Fibers use Noradrenalin and Norepinephrine.

AUTONOMIC NERVOUS SYSTEM AND DISEASE/ILLNESS

Stress and Heart Disease

- Individuals with Chronic Stress are more prone to Heart Disease. Stress activates the Sympathetic Nervous System.
- This causes Constriction of the Blood Vessels, Increases Heart Rate, and Heightens Cholesterol Production.

Stress and Healing

- Stress promotes Sympathetic Nervous System Activity.
- This causes decreased Repair of Cell Structures.
- Parasympathetic Nervous System Activity promotes Homeostasis and facilitates Cellular Repair and Tissue Restoration.
- Parasympathetic Nervous System Activity is heightened during Restorative Sleep—Stage 4 Sleep.
- Fibromyalgia—an illness characterized by muscular and joint pain—is believed to be caused in part by a lack of Stage 4 Restorative Sleep.

Denervation

- Denervation is an interruption of Neuronal Innervation.
- Causes Hypersensitivity—the injured neuron experiences hypersensitivity to its own neurotransmitter.
- This results in Dysregulation of the ANS.

Symptoms of Autonomic Dysregulation
- Cold, dry, flaky skin with loss of hair over the denervated skin region.
- Alterations in sweating.
- The nails may become thickened.
- The denervated skin region may become smooth and glossy.

Horner's Syndrome
- Horner's Syndrome is an ANS disorder.
- Results from a transection of the Oculomotor Sympathetic Pathway (Cranial Nerve 3).
- Causes:
    - Ipsilateral Mioses (constriction of the pupil)
    - Partial Ptosis (drooping of the ipsilateral eyelid)
    - Flushed Dry Skin on the Ipsilateral Face
    - Ipsilateral Sunken Eyeball

AUTONOMIC NERVOUS SYSTEM CHARACTERISTICS

SYMPATHETIC

- Activated in Short Bursts When Individual is Stressed, Angry, Frightened, Excited. May be Activated Continuously if Individual has Chronic Stress
- Fight/Flight Response
- Pupils Dilate
- Increased Focus on Event/Environment
- Heart Rate Increases
- Blood Pressure Increases
- Respiration Increases and Quickens
- Vasoconstriction of Arteries to Bring Blood to Heart Faster
- Saliva Thickens
- Blood is Diverted Away from the GI Tract
- Digestive Juices Stop Production
- Peristalsis Stops

- Activation of all Muscle Groups
- Cell Destruction
- Body Temperature Changes

PARASYMPATHETIC

- Activated Continuously to Maintain Homeostasis of Body Systems. Will Shut Down When Sympathetic Nervous System Takes Over

- Homeostasis
- Pupils Constrict
- Decreased Awareness of Environment
- Heart Rate Slows
- Blood Pressure Decreases
- Respiration Slows and Becomes Shallow
- Vasodilation of Arteries

- Saliva Thins
- Blood Returns to Viscera (in GI Tract)
- Digestive Juices Begin to be Secreted Again
- Peristalsis is Continuous Unless Sympathetic Nervous System Shuts Parasympathetic Nervous System Down
- Relaxation of Most Muscle Groups
- Cell Repair
- Body Temperature is Maintained at a Constant Level (98.6º F)

Enteric Nervous System

The ENS is an Independent Circuit that is loosely connected to the Central Nervous System (CNS) but can function alone without instruction from the CNS.

LOCATION

Located in sheaths of tissue that line the Esophagus, Stomach, Small Intestine, and Colon.

COMPOSITION

- Composed of a Network of Neurons, Neurotransmitters, and Proteins.
- The ENS contains 100 Million Neurons—more than the Spinal Cord.

CHEMICAL SUBSTANCES IN THE BRAIN AND ENTERIC NERVOUS SYSTEM

- Every Chemical Substance that helps to control the Brain has been found in the Intestines:
  - Major Neurotransmitters such as Serotonin, Dopamine, Glutamate, Norepinephrine, and Nitric Oxide
  - Two Dozen small Brain Proteins called Neuropeptides
  - Mast Cells—cells of the immune system
  - Enkephalins—one class of the body's natural opiates
  - Benzodiazepines—the family of psychoactive chemicals that include Valium and Xanax

DEVELOPMENT OF THE ENTERIC NERVOUS SYSTEM

- A formation of tissue called the Neural Crest develops early in Embryogenesis.
- One section of the Neural Crest turns into the CNS (Brain and Spinal Cord).
- Another piece migrates to become the ENS.
- Only later in fetal development are the two systems connected via the Vagus Nerve (CN 10).

COMMUNICATION BETWEEN THE CENTRAL AND ENTERIC NERVOUS SYSTEMS

- The CNS sends signals to the ENS through a small number of Command Neurons.
- These Command Neurons then send signals to the ENS's Interneurons (small neurons that connect the two major neurons).
- Both Command Neurons and Interneurons are located in two layers of Intestinal Tissue called the:
  - Myenteric Plexus
  - Submucosal Plexus
- Command Neurons appear to control the Pattern of Activity in the Gastrointestinal (GI) Tract. The Command Neurons are an Independent System of the ENS.
- The CNS appears to primarily control the Firing Rate of the Command Neurons.
- The Vagus Nerve only Alters the Firing Rate of the Command Neurons.

THE MYENTERIC AND SUBMUCOSAL PLEXES

- The Plexes have Sensors for Sugar, Protein, and Acidity Levels.
- These Sensors Monitor the Progress of Digestion and determine how the Intestines should mix and propel its contents.

DRUG INTERACTIONS AND THE CONNECTION BETWEEN THE ENTERIC AND CENTRAL NERVOUS SYSTEMS

- When pharmacologists design a drug to have effects on the brain, it very likely has concomitant effects on the GI Tract that are undesired—side effects.
- The Intestine contains a substantial amount of Serotonin. When Pressure Receptors in the GI Tract's lining are stimulated, Serotonin is released and begins the reflexive motion of Peristalsis.
- Twenty-five percent or more of individuals taking a Selective Serotonin Re-uptake Inhibitor (SSRI—a class of antidepressants including Prozac, Paxil, and Zoloft) experience accompanying GI problems—nausea, diarrhea, and constipation.
- These drugs act on Serotonin, preventing its re-uptake by receptor cells. It is beneficial if higher levels of Serotonin remain in the CNS.
- But when high levels of Serotonin are released and remain in the GI Tract, it causes bowel problems.
- The SSRIs also double the speed at which food is passed through the Colon, explaining why some people taking SSRIs experience diarrhea.
- Sometimes, too much antidepressant drugs can have the effect of producing constipation.
- Some Antibiotics—like Erythromycin—act on the GI Receptors to produce Oscillations, causing cramping and nausea.
- Drugs like Morphine and Heroin attach to the intestine's Opiate Receptors and can produce constipation.
- The ENS can become addicted to drugs just like the CNS does.
- People with Alzheimer's and Parkinson's disease often experience constipation, because the pathology of their CNS disorder also causes Dysregulation of the Intestine's functioning.

SIMILARITIES BETWEEN THE CENTRAL AND ENTERIC NERVOUS SYSTEMS

- Both the CNS and ENS act similarly when deprived of input from the external world.
- During sleep, the CNS produces 90-minute cycles of Slow Wave Sleep punctuated by periods of Rapid Eye Movement (REM) Sleep in which dreams occur.
- During the night when the ENS has no food, it produces 90-minute cycles of Slow Wave Muscle Contractions punctuated by short bursts of Rapid Muscle Movements.
- The two systems influence each other while in this state. Patients with bowel problems have been shown to have abnormal REM Sleep.
- This finding seems consistent with the folk wisdom that indigestion can cause nightmares.

ENTERIC AND CENTRAL NERVOUS SYSTEMS RESPONSE TO FIGHT/FLIGHT SITUATIONS

- When individuals encounter frightening situations, the CNS releases Stress Hormones that prepare the body to fight or flee (via the Sympathetic Nervous System).
- The GI Tract also contains many sensory nerves that are stimulated by this chemical surge. These produce the common sensation of butterflies in the stomach.
- The CNS instructs the GI Tract to shut down during Fight/Flight situations.
- Fear and Chronic Stress also cause the Vagus Nerve to increase the firing rate of Serotonin Circuits in the GI Tract.
- When the GI Tract becomes over-stimulated with high levels of Serotonin, bowel problems can occur (eg, colitis, irritable bowel syndrome).
- Similarly, when nerves in the Esophagus are highly stimulated by an increase in the release or production of Serotonin or Norepinephine, the Esophagus reacts by constricting—thus making swallowing difficult. This is probably where the phrase "choked up with emotion" is derived from.

Pain

PAIN

- Pain is the sensory experience that is unpleasant and is associated with possible tissue damage.
- The detection of pain indicates that a pathological condition may be occurring in the organism.
- The detection of pain is called Nociception.
- Nociceptors are Receptors that detect harmful stimuli.

TYPES OF PAIN

Somatic Pain

- Somatic Pain occurs from the body (Skin, Skeletal Muscles, Bones).
  - Superficial Somatic Pain: usually Well-Localized (eg, Pin Prick).
  - Deep Somatic Pain: usually NOT Well-Localized (eg, Muscular Ache).

Visceral Pain

- Visceral Pain occurs from the Viscera (Internal Organs, Glands, Smooth Muscle).
- Visceral Pain is Dull or Diffuse and NOT Well-Localized.
- Usually accompanied by an Autonomic Nervous System Response (change in Heart Rate and Blood Pressure, Nausea).

Qualities of Pain

- Dull Ache—tends to be Diffuse.
  - Dull Aching Pain tends to last a long time because it is carried by slow conducting, small, unmyelinated, C Fibers.
- Sharp Pain—tends to be Well-Localized
  - Sharp Pain tends to last a short time because sharp pain is carried by fast conducting large A Delta Fibers.

Pain Receptors

- Pain Receptors are believed to be Free Nerve Endings.
- If stimulated intensely enough, other types of receptors may act as pain receptors as well.

MAJOR PAIN PATHWAYS

Spinothalamic Spinal Cord Tracts

- The Spinothalamic Tracts are ascending somatic sensory pathways that receive pain information from the skin and skeletal muscles.
- Sensory Nerves carry pain information from the Skin and Skeletal Muscles in the periphery to the Dorsal Horn of the SC.
- When these Spinal Nerves synapse in the Dorsal Horn they release a Neurotransmitter called Substance P.

- The Spinothalamic tracts travel from the SC to the Thalamus and send projections to the Cortex for conscious pain detection and interpretation.
- Substance P is transmitted via the Spinothalamic Tracts to the Thalamus and Cortex.

Spinoreticular Spinal Cord Tracts
- The Spinoreticular Tracts are descending sensory tracts that receive pain information from the periphery through afferent spinal nerves that synapse in the Reticular Formation of the Brainstem.
- The Spinoreticular Tracts have their origin in the Medullary Reticular Formation.
- There they travel to the Raphe Nuclei of the Brainstem. The Raphe Nuclei are a group of nuclei located in the Medulla. They are situated along the midline.
- When the Raphe Nuclei become excited they release Endorphins through a descending pathway to the place of pain origin to decrease the pain sensation.

Trigeminothalamic Tracts
- The Trigeminothalamic Tracts have their origin in the Trigeminal Lemniscus in the Brainstem.
- The Trigeminal Lemniscus projects afferent fibers from the Trigeminal Nerve (CN 5) to the Thalamus and then to the Cortex.
- This tract carries pain sensation from the face.

Pathways to the Cortex for the Conscious Detection of Pain
- Pain messages from the SC Tracts are projected to the Thalamus and then to the Cortex for Conscious Detection and Interpretation.
- The Primary Somatosensory Area (SS1), located in the Postcentral Gyrus, detects incoming somatosensory data from the periphery.
- Pain messages are then projected to the Secondary Somatosensory Area (SS2) for Interpretation.
- SS2 is located just posterior to SS1.
- Pain messages are then projected to the Posterior Multimodal Association Area for the integration of pain information with other sensory data.
- Example: The Multimodal Association Area integrates pain with smell. The individual will learn to associate the smell of spoiled food with abdominal pain—in order to refrain from eating spoiled food in the future.
- The Multimodal Association Area sends projections to the Limbic System for the integration of sensation, emotion, and memory.
- The individual is able to remember the smell of spoiled food at later dates, and will remember the pain associated with eating spoiled food.

HOW THE BODY CONTROLS PAIN

Gate Control Theory
- One of the first theories of Pain Control was the Gate Theory, proposed by Melzak and Wall in 1965.
- Simplistically, the Gate Theory suggested that the transmission of pain information is blocked in the Dorsal Horn, closing the gate to pain.
- Lamina II, or the Substansia Gelantinosa in the Dorsal Horn, was suggested as the site of interference with pain message transmission.
- The Gate Theory suggested that Afferent Sensory Fibers carry pain sensation from the periphery into the Dorsal Horn of the SC.
- This information synapses in the Substantia Gelatinosa.
- T-Cells are specialized cells in the SC. They begin to fire and cause the release of Substance P—a neurotransmitter involved in the sensation of pain.
- If the Substantia Gelantinosa can be facilitated by another pathway (a collateral pathway), the T-Cell firing will diminish causing a decrease in pain transmission.
- Much of the Gate Theory has been disproved.

The Counterirritant Theory of Pain Cessation
- The Counterirritant Theory of Pain Cessation has incorporated research-based findings first suggested by the Gate Theory.

- The Counterirritant Theory suggests that Non-Nociceptors in the Dorsal Horn inhibit the excited Nocireceptors (also in the Dorsal Horn).
- For example, Pressure (like rubbing the painful area) stimulates Mechanoreceptive Afferent Fibers.
- The proximal branches of the Mechanoreceptors in the Dorsal Horn activate Interneurons that synapse on the excited Nociceptors (in the Dorsal Horn).
- These Interneurons release the Neurotransmitter Enkephalin—a chemical in the family of Endorphins.
- Enkephalin binds with the excited Nociceptor and diminishes the release of Substance P.
- Enkephalin binding on the Nociceptor inhibits the transmission of Nociceptive Signals, thus decreasing the sensation of pain.

ANALGESIC INHIBITION OF PAIN

- Analgesia is an absence of pain in response to stimulation that would otherwise cause pain.
- Analgesic Mechanisms can be activated by:
  - Endorphins—naturally occurring substances that diminish the sensation of pain
  - Pharmaceuticals that diminish the sensation of pain
- Endorphins include Enkephalin, Dynorphin, and Beta-Endorphin.
- Opiates are the family of Analgesic Drugs that block Nociceptor signals without affecting other sensations.
- Both Endorphins and Analgesic Drugs bind to the same receptor site.
- The Inhibition of Nociceptive Information can also be inhibited by the Supraspinal Levels of the Nervous System. These are the Brainstem Centers that provide natural Analgesia. They are referred to as Pain Inhibiting Centers:
  - The Raphe Nuclei in the Medulla
  - The Periaqueductal Gray in the Midbrain
  - The Locus Ceruleus in the Pons
- When the Raphe Nuclei are stimulated, axons projecting to the SC release the Neurotransmitter Serotonin in the Dorsal Horn. This release of Serotonin inhibits the transmission of Nociceptive Signals.
- When the Periaqueductal Gray is stimulated, it also produces an analgesic effect by activating the Raphe Nuclei.
- The Ceruleospinal Tract originates at the Locus Ceruleus in the Pons. When stimulated, it causes Norepinephrine to bind to the Spinothalamic Tract in the Dorsal Horn. Binding of Norepinephrine to the Spinothalamic Tract suppresses the release of Substance P, thus diminishing pain messages to the Cortex.
- Narcotic Drugs (derived from opium or opium-like compounds) bind to receptor sites in the Periaqueductal Gray, the Raphe Nuclei, and the Dorsal Horn. By binding to these receptor sites, Narcotic Drugs induce analgesia and stupor (a state of reduced consciousness).

STRESS-INDUCED ANALGESIA

- The Pain Inhibiting Centers can be activated Naturally by Injury and Athletic Over-Exertion.
- Often, people injured during accidents, disasters, or athletic competition may not feel pain until the event has passed.
- Stress occurring during the event may trigger the Pain Inhibition Centers.
- Stress-Induced Analgesia involves activation of:
  - The Raphe Nuclei Descending Tracts
  - The Release of the Hormonal Endorphins from the Pituitary Gland (particularly Beta Endorphins)
  - The Release of Hormonal Endorphins from the Adrenal Medulla (particularly Enkephalins)
- The Hormonal Endorphins bind to the Opiate Receptors in the Brain and SC.
- Beta-Endorphins are the most Potent Endorphins and can trigger analgesic affects that last for hours.

PAIN TRANSMISSION CAN BE DIMINISHED AT SEVERAL NERVOUS SYSTEM LEVELS

The Periphery

- Non-Narcotic Analgesics (eg, aspirin) decrease the synthesis of Prostaglandins, thus preventing Prostaglandins from sensitizing peripheral pain receptors. Prostaglandins are a large group of biologically activated, carbon-20, unsaturated fatty acids.

Dorsal Horn
- Inhibitory Neurons in the Dorsal Horn release Enkephalin or Dynorphin. These can diminish pain sensation through interneurons that bind to the excited Nociceptor. This is the principle of the Counterirritant Theory.

Supraspinal Descending Systems
- The Raphe Nuclei, Periaqueductal gray, and the Locus Ceruleus can inhibit Nociceptive Information.

Hormonal System
- Involves the release of the Hormonal Endorphins from the Pituitary Gland (particularly Beta-Endorphins).
- Involves the release of Hormonal Endorphins from the Adrenal Medulla (particularly Enkephalins).

Cortical Level
- The Cortical Detection and Interpretation of Pain can be altered by an individual's expectations, distraction level, anxiety, and belief (particularly regarding placebo effects).

Pain Transmission Can Also Be Intensified at Several Nervous System Levels

- Edema and Endogenous Chemicals can sensitize Free Nerve Endings in the periphery.
- For example, following a minor burn injury, sensory stimuli that would normally be innocuous can cause heightened pain.
- Fear and Anxiety can also heighten the experience of pain.

Pain Tolerance and Pain Thresholds

- Pain Thresholds are related to the sensitization of one's peripheral Nociceptors.
- Some individuals may have naturally heightened Nociceptors compared to others.
- Pain Thresholds are also intimately related to past experience, expectations, and fear.
- Women who give birth a second time often report that the second birthing experience was not as painful; labor tends to be shorter. This may relate to experiencing less anxiety and fear during the second birth.

Referred Pain

- Referred Pain is pain that is perceived to originate from one body region, when it actually originates from a different body region.
- Usually Referred Pain occurs when Visceral Pain (from an internal organ or gland) is perceived as originating from a Somatic Area (such as the skin or skeletal muscles).
- For example, during a heart attack, the brain may misinterpret the nociceptive information as arising from the skin on the medial left arm.
- Similarly, gallbladder pain is often referred to the right subscapular region.
- The phenomenon of Referred Pain can be explained by the Dermatomal Distribution:
- In a heart attack, Nociceptive Information from the heart projects to and from the SC segment T1. The dermatomal sensation of the medial left arm also projects to and from T1.
- Because the Cortex is unfamiliar with pain messages received from the heart, it initially interprets the pain as originating from the left arm—until the pain becomes excruciating.
- Some dorsal root neurons have two peripheral axons—one that innervates the skin and skeletal muscle, and one that innervates the viscera. Stimulation of the visceral branch of a dual receptive neuron may be the source of cortical misinterpretation.

Techniques to Reduce Pain

Stimulation of the Mechanoreceptors (Massage)
- Stimulation of the Mechanoreceptors is based on the Counterirritant Theory and may involve rubbing or massaging the painful area.

- The proximal branches of the Mechanoreceptors in the Dorsal Horn activate Interneurons that synapse on the excited Nociceptors (also in the Dorsal Horn).
- These interneurons release the neurotransmitter Enkephalin—which binds with the excited Nociceptor and diminishes the release of Substance P (released from the excited Nociceptors).
- Enkephalin Binding on the Nociceptor inhibits the transmission of Nociceptive signals, thus decreasing the sensation of pain.

TENS (Transcutaneous Electrical Nerve Stimulation)
- Also based on the Counterirritant Theory
- The therapist finds the dermatomal region of the pain area and applies TENS to activate the Mechanoreceptors.
- Technique has marginal effectiveness.

Ultrasound
- Involves conversion heating; as ultrasound is propagated through tissue it is absorbed and converted into heat.
- Ultrasound is one of the most effective heat modalities for deep structures; reported depths of penetration travel 5 to 6 cm below the superficial layers of the skin.

Hot Packs
- Hot packs contain a silicone gel that, when immersed in hot water, has the capacity to absorb and hold a great amount of heat.
- Hot packs are not able to provide deeply penetrating heat, as is ultrasound.
- Recommended for large body areas, such as the back.
- Provides temporary relief of pain.

Cold Packs
- Cold packs are useful for anesthetizing sensory receptors to relieve pain.
- The use of cold can also decrease inflammation, which can relieve pain.

Paraffin
- Heated wax; a method of superficial heat conduction.
- Recommended for specific joint tightness and pain—particularly in the hands.

Whirlpool Hydrotherapy
- A method of convection heat using water.
- Particularly useful for large body areas such as the back.

Fluidotherapy
- Much like a dry whirlpool with ground corn husk particles instead of water as the heating medium.
- The particles are circulated by hot air blown within the fluidotherapy machine.
- Useful for pain relief in the upper extremities—particularly the hand, wrist, and forearm.

Acupuncture
- The mechanism by which Acupuncture works is unknown.
- Most likely involves the release of Endorphins.
- Can effectively provide lasting pain relief.

Dorsal Rhizotomy
- Involves cutting selected dorsal roots or the Spinothalamic Tracts to eliminate pain.
- Poor results.

Stress Management and Meditation
- Involves learning how to naturally release Endorphins.
- Learning how to lower pain thresholds by using visual imagery.

Biofeedback
- Involves learning to release tension in muscles that may be tensed because the individual is using those muscles to guard or protect the painful area.
- Can offer patients an effective pain management strategy.

Fibromyalgia

- Fibromyalgia is a Chronic Pain Syndrome that does not have a clear etiology.
- It is believed that mental and/or physical stress, trauma, and lack of sufficient restorative sleep periods contribute to the disorder.
- Restorative sleep is Deep Stage 4 Sleep-the period of Rapid Eye Movement (REM) Sleep.
- Humans generally experience two 90 minute cycles of Deep Stage 4 Sleep.
- It is in Stage 4 Sleep that cellular repair is believed to be most efficient.
- Individuals who are deficient in Deep Stage 4 Sleep, over time, appear to develop the symptoms of Fibromyalgia.
- Symptoms include tenderness of muscles and adjacent soft tissues, stiffness of muscles, and aching pain. Multiple tender points can be found on palpation.
- The painful area follows a regional rather than a dermatomal or peripheral nerve distribution.
- Increased restorative sleep and appropriate exercise have been found to benefit people with Fibromyalgia.

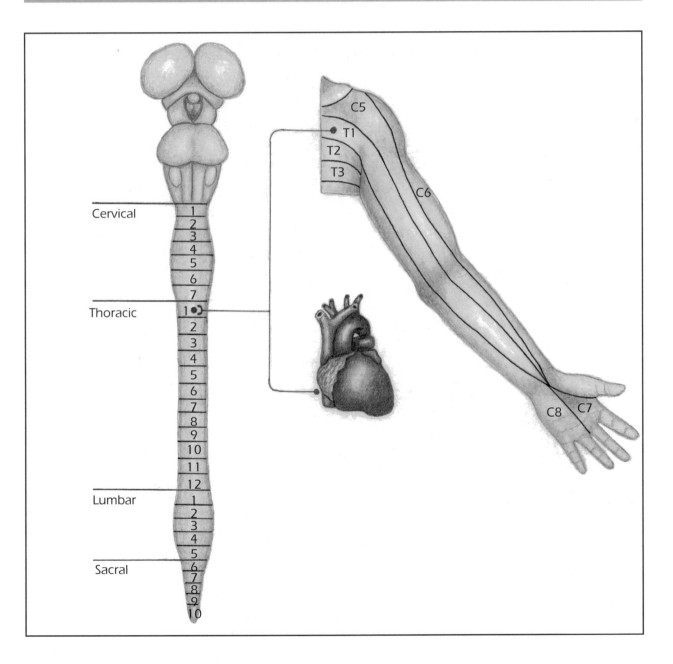

Peripheral Nerve Injury and Regeneration

PERIPHERAL NERVE INJURY AND REGENERATION

Neuropathy
- Neuropathy is a general term for pathology involving One or More Peripheral Nerves.

Dermatomal Distribution
- The Dermatomes are Skin Segments that are innervated by Specific Peripheral Nerves.
- The Dermatomal Skin Distribution corresponds closely to the Skeletal Muscles that are innervated by a Specific Peripheral Nerve.
- When a therapist evaluates Sensation, the therapist will test each Dermatomal Skin Segment to identify possible Loss of Sensation.
- Identification of a Specific Dermatomal Skin Segment that has Lost Sensation will indicate the Lesion Level.
- For example, a Loss of Sensation on the Lateral Forearm and the Lateral Hand may result from C6 Peripheral Nerve Loss.

Complete Severance of a Peripheral Nerve
- A Completely Severed Peripheral Nerve results in a Loss of Sensation, Loss of Motor Control, and a Loss of Reflexes in the structures innervated by a Specific Peripheral Nerve.

Nerve Compression
- Nerve Compression results in a Loss of Proprioception and Discriminative Touch.
- Pain and Temperature initially Remain Intact.
- This occurs because Compression of a Nerve affects the Large Myelinated Fibers first—these are the Fibers that carry Proprioceptive and Discriminative Touch Information.
- There is Initial relative Sparing of the Smaller Fibers that carry Pain and Temperature.
- When Compression of a nerve occurs, Sensory Loss proceeds in the following order:
  - Conscious Proprioception and Discriminative Touch
  - Cold
  - Fast Pain or Sharp Pain
  - Heat
  - Slow Pain or Dull, Diffuse, Aching Pain

When Compression Resolves
- When the Compression is relieved, Abnormal Sensations called Paresthesias occur as the Blood Supply increases.
- Paresthesias include Burning, Pricking, and Tingling Sensations.
- After Compression is resolved, Sensation returns in the Reverse Order in which it was lost:
  - Slow Pain or Dull, Diffuse, Aching Pain

- Heat
- Fast Pain or Sharp Stinging Sensations
- Cold
- Conscious Proprioception and Discriminative Touch
- This Process can be observed when one of the body's limbs "falls asleep" as a result of Nerve Compression.

Flaccidity

- When a Peripheral Nerve is Injured, the Muscles that are innervated by that nerve become Flaccid and gradually Atrophy.
- Paralysis and Loss of Sensation occur just Distal to the Lesion.

Schwann Cells

- Schwann Cells compose the Myelin in the Peripheral Nervous System (PNS).
- Schwann Cells Produce Nerve Growth Factor (NGF).
- This allows Peripheral Nerve Damage to resolve (unlike damage in the central nervous system, or CNS).
- Nerve Regrowth occurs approximately 1 mm per day or 1 inch per month.
- A C6 Peripheral Nerve Injury in the Upper Arm could take a year or more to fully regenerate.
- Compression Injuries recover more quickly than Complete Severance of a Nerve.

TERMINOLOGY OF SENSORY PATHOLOGY

Hypoesthesia: A Decrease in Sensory Perception.
Hyperesthesia: An Increase in Sensory Perception, eg, Heightened Perception of Pain and Temperature.
Paresthesia: The Occurrence of Unusual Feelings, such as Pins and Needles.
Dysesthesia: Unpleasant Sensation such as Burning, *Causalgia* is an Intense Burning Pain accompanied by Trophic Skin Changes.

Autonomic Dysfunction

- Autonomic Dysfunction may result in Vasomotor Nerve Disturbances.
- Vasomotor Nerves are nerves that have Muscular Control over the Blood Vessel Walls. They control Vasodilation and Constriction of Blood Vessels.
- Autonomic Trophic Changes (also called Nutritional Changes) affect the skin, hair, and nails. The Skin may become Smooth and Glossy, the Nails can become Thickened, and the Hair on the Denervated Skin Area Thins or Falls Out.

Brachial Plexus

- The Brachial Plexus is a Network of Peripheral Spinal Nerves including C5, C6, C7, C8, and T1.
- The nerves that Supply the Upper Extremities emerge from the Brachial Plexus.
- The Brachial Plexus is a Site of Common Compression Syndromes:
  - Radial Nerve Compression: Wrist and Finger Drop, Tennis Elbow, and Saturday Night Palsy
  - Medial Nerve Compression: Carpal Tunnel
  - Ulnar Nerve Compression: Clawhand Deformity and Deformity of the 4th and 5th Digits

Lumbar Plexus

- The Lumbar Plexus is a Network of Peripheral Spinal Nerves formed by L1 through L4.
- Also a Common Site of Compression Syndromes:
  - Sciatic Nerve Compression: Sciatica
  - Peroneal Nerve Compression: Foot Drop

TYPES OF PERIPHERAL NEUROPATHY

Mononeuropathy

- Mononeuropathy involves damage to a Single Nerve.
- Usually due to Compression or Entrapment.
- Wrist Drop: Radial Nerve Entrapment
- Foot Drop: Peroneal Nerve Entrapment

Radiculopathy

- Nerve Root Impingement.
- Results from a Lesion affecting the Dorsal or Ventral Roots.
- Can result from Herniated Vertebral Discs.

Plexopathy

- Caused by Damage to One of the Plexes—Brachial or Lumbar.
- Involves Multiple Peripheral Nerve Damage.

Polyneuropathy

- Usually involves Bilateral Damage to More than One Peripheral Nerve.
- Example: STOCKING and GLOVE Polyneuropathy. Usually caused by a Disease Process such as Diabetes.
- With Mild Disease, only the Distal Lower Extremities are involved.
- With More Severe Disease, the Distal Upper Extremities are also involved.

DEEP TENDON REFLEXES

- When a Peripheral Neuropathy occurs, the individual often presents with an Absence of Distal Deep Tendon Reflexes.
- An Asymmetric Decrease in Deep Tendon Reflexes occurs in:
- Radiculopathy (Nerve Root Impingement on One Side)
- Plexopathy (Brachial or Lumbar Plexus Involvement)
- Mononeuropathy (Unilateral Spinal Nerve Injury)

Neuropathies Caused by a Disease Process

Guillain-Barré

- Guillain-Barré is an Acute Inflammatory Polyradiculopathy.
- It is preceded by an infectious illness that usually clears before the neurologic dysfunction becomes evident.
- Involves progressive ascending weakness (as in Stocking and Glove neuropathy).
- Involves complete Tendon Areflexia.
- Although recovery may be slow, progress is one of almost complete improvement.

Diabetes Mellitus

- Polyneuropathies, Mononeuropathies, Plexopathies and Autonomic Neuropathies can ALL be found in Diabetes.
- Diabetes can be associated with abnormalities at any level of the PNS.
- Autonomic neuropathies involve the gastrointestinal, cardiovascular, and genitourinary systems.
- Autonomic neuropathy can involve postural hypotension (when one rises from a horizontal position, blood pressure may drop to precariously low levels), diarrhea, impotence, urinary retention, and increased sweating.

Stocking and Glove Syndrome

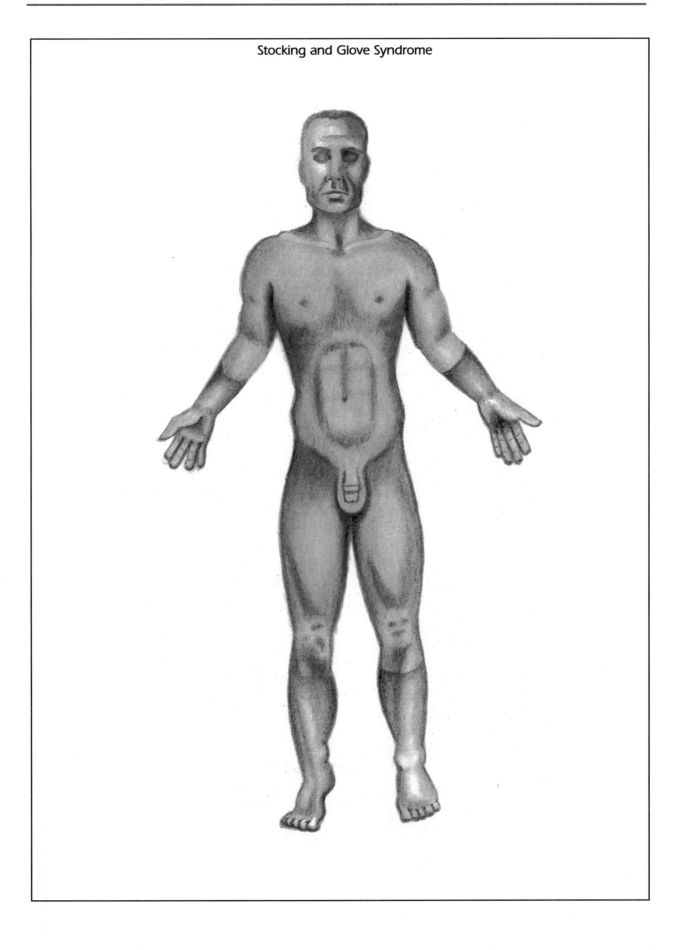

Severed Axon

- When an Axon is severed, the Portion connected to the Cell Body is referred to as the Proximal Segment.
- The Portion that is now disconnected from the Cell Body is called the Distal Segment.

Leakage of Protoplasm

- Immediately after injury, Protoplasm leaks out of each severed end and the two segments retract away from each other.

Wallerian Degeneration

- The Distal Segment undergoes a process called Wallerian Degeneration.
- The Myelin Sheath pulls away from the Distal Segment.
- The Distal Segment swells and breaks into smaller segments.
- The Axon Terminals degenerate rapidly.
- The entire Distal Segment dies.
- Glial Cells scavenge the area and clean up the debris from the degenerated Distal Segment.

Central Chromatolysis

- Sometimes the Cell Body of the Proximal Segment undergoes Degenerative Changes called Central Chromatolysis. This may lead to Cell Death.
- Simultaneously, the Postsynaptic Neuron—no longer innervated by the Presynaptic Neuron—may also degenerate and die.

Sprouting

- The Regrowth of damaged Axons is called Sprouting.
- Sprouting takes two forms:
- Collateral
- Regenerative

Collateral Sprouting

- Collateral Sprouting occurs when a denervated Postsynaptic Neuron is Re-Innervated by Branches of an Intact Axon located near the Damaged Axon.
- In other words, another Neuron projects a Collateral Axon branch to the Cell Body of the Postsynaptic Neuron.

Regenerative Sprouting

- Regenerative Sprouting occurs when an Axon and its Postsynaptic Neuron have both been Damaged.
- The Proximal Segment of the Damaged Neuron projects Side Sprouts to form New Synapses with other Undamaged Postsynaptic Neurons.

Functional Regeneration Occurs in the Peripheral Nervous System

- Regeneration of Axons occurs primarily in the PNS.
- This is partly due to the Production of NGF by the Schwann Cells.
- The Schwann Cells are the Myelin around the PNS Axons; Schwann Cells do NOT exist in the CNS.
- Oligodendrites are the Myelin Cells that wrap around the CNS axons; Oligodendrites do NOT produce NGF.
- There is little or no recovery in the CNS.

Recovery Speed in the Periphery

- Regeneration of Damaged Axons in the PNS is slow—1 mm of growth per day or 1 inch of recovery per month.
- A Severed Nerve in the Upper Arm—such as C6—will take at least a year to recover.

Problems that Can Occur in the Nerve Regeneration Process

- Sometimes an Axon will Innervate a New Postsynaptic neuron that is inappropriate.
- For example, after injury of a peripheral nerve, Motor Axons may innervate Different Muscles than before the injury.
- When the Neuron fires, this results in Unintended Movements called Synkineses.
- Synkineses are usually short-lived, as the person Relearns Muscle Control.
- Similarly, in the sensory systems, innervation of a Sensory Receptor by Axons that previously innervated a different type of Sensory Receptor, can cause confusion of sensory modalities.

SYNAPTIC CHANGES AFTER INJURY

Synaptic Effectiveness
- Local Edema in the injured area causes Compression of the Presynaptic Neuron's Cell Body or Axon.
- When the Edema is resolved Synaptic Effectiveness returns.

Denervation Hypersensitivity
- Denervation Hypersensitivity occurs when the Postsynaptic Neuron—of an Injured Presynaptic Neuron— becomes Hypersensitized to its own Neurotransmitter.
- This occurs because the Postsynaptic Neuron develops New Receptor Sites that can respond to Neurotransmitters released by other nearby neurons.
- This occurs only for a short time.
- Produces Temporary Muscle Twitches and Pain.

Synaptic Hypereffectiveness
- Synaptic Hypereffectiveness occurs when only some Branches of a Presynaptic Axon are Destroyed leaving the Majority of Branches intact.
- The Remaining Axon Branches receive all of the Neurotransmitter Substance that would normally be shared among only the Terminal End Branches.
- This results in a larger than normal amount of neurotransmitter being released onto Postsynaptic Receptors.

Unmasking of Silent Synapses
- In the normal PNS many synapses seem to be unused unless other pathways become injured.
- Unmasking of Silent Synapses occurs when Previously Non-Used Synapses become active after other pathways have been damaged.

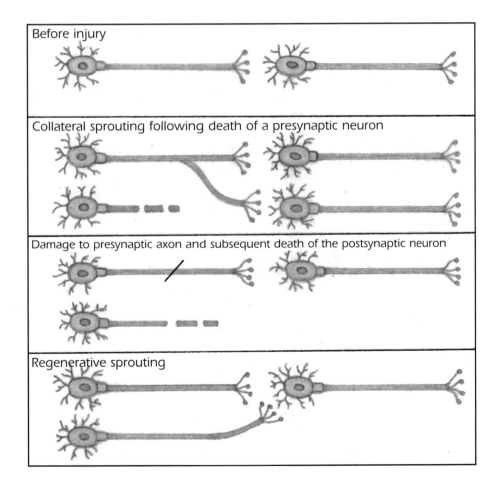

Before injury

Collateral sprouting following death of a presynaptic neuron

Damage to presynaptic axon and subsequent death of the postsynaptic neuron

Regenerative sprouting

SECTION 17

Phantom Limb Phenomenon

PHANTOM LIMB PHENOMENON

- Phantom Limb Phenomenon is the Sensation that an Amputated Body Part still remains.
- If the Sensation is Painful, it is referred to as Phantom Pain.

Phantom Limb

- Any Non-Painful Sensation in the Amputated Limb.

Phantom Pain

- Painful Sensations that seem to occur in the Lost Limb.

Stump Pain

- Painful Sensations localized to the Stump of an Amputated Body Part.

Phantom Pain

- Phantom Pain is Cortically Perceived—there is NO Peripheral Component.
- Phantom Pain usually cannot be ended by Nerve Blocks.
- Phantom Pain is described as Excruciating, Sticking, Cramping, Burning, and Squeezing.

Phantom Pain and Phantom Limb Are Central Nervous System Phenomena

- Phantom Pain is not mediated by Peripheral Nerve Signals, as is Stump Pain.
- Phantom Pain and Phantom Limb Sensations are largely due to CNS Phenomena.
- The Cortical Map of the Body (the Sensory and Motor Homunculi) still retain the Anatomical Image of the Amputated Body Part.
- The Brain believes that the Amputated Body Part still remains.
- On Transcranial Magnetic Stimulation, No Change can be observed in the Cortical Map Area after the Amputation.

Phantom Sensations Lessen Over Time

- Other Areas of the Cortical Map eventually appropriate the Cortical Region that once mediated Sensation/Movement of the Amputated Limb.
- At this time, Transcranial Magnetic Stimulation shows changes in the Cortical Map.

Phantom Limb Sensation Can Aid in Gait Training

- Patients with Phantom Sensations—that are NOT painful—report that they can use the Phantom Image to propel their Prosthesis.
- Phantom Pain interferes with Gait Training.

Treatment of Phantom Pain

- Treatment for Phantom Pain is often ineffective.
- TENS works to a moderate degree.

- Vibration to the Stump may work temporarily.
- Analgesics or Painkillers work to a moderate degree, and only temporarily.

Phantom Limb Sensation Can Be Experienced in Spinal Cord Injuries

- The Phantom Limb Sensation that people with SCI experience tends to resolve quickly.
- Other areas of the Cortical Map appropriate the areas once used to mediate the paralyzed limbs.
- This may relate to the fact that in SCI, NO Peripheral Stimulation can continue, as it might in an amputee who still retains some portion of the limb.

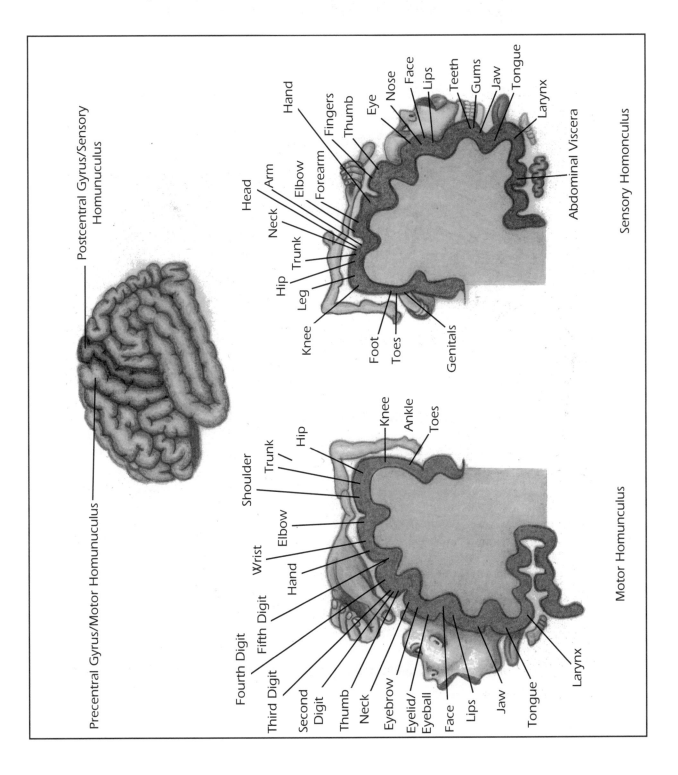

Spinal Cord Tracts

REVIEW OF SPINAL CORD ANATOMY

- 31 Pairs of Spinal Nerves:
  - 8 Cervical, 12 Thoracic, 5 Lumbar, 5 Sacral, and 1 Coxygeal

ASCENDING SPINAL NERVES

- Ascending Spinal Nerves carry SENSORY Information from the Periphery (beginning with a Sensory Receptor) to the Spinal Cord.
- The Sensory Spinal Nerves enter the SC through the Dorsal Root and Rootlets (in the PNS).
- Once in the Dorsal Horn (in the CNS) the Spinal Nerves Synapse with an Interneuron.
- The Interneuron Synapses with an Ascending Sensory Spinal Cord Tract (in the CNS).

ASCENDING SPINAL CORD TRACTS

- Ascending Spinal Cord Tracts carry the Sensory Information up the SC, to the Brainstem, Thalamus, and Cortex.
- Some Sensory SC Tracts carry Sensory Information from the Brainstem to the Cerebellum.

DESCENDING SPINAL CORD TRACTS

- Descending SC Tracts carry MOTOR Information from the Cortex, through the Internal Capsule, through the Thalamus, Brainstem, and to the SC (in the CNS).
- Some Motor Tracts originate in the Cerebellum and Brainstem.
- When Descending Motor SC Tracts are ready to Exit the Cord, they Synapse on an Interneuron in the Ventral Horn of the SC.
- The Interneuron then synapses on a Motor Neuron in the Ventral Horn (in the PNS).
- The Motor Neuron in the Ventral Horn Synapses on a Descending Motor Spinal Nerve and Exits the Cord through the Ventral Rootlets and Root (in the PNS).
- The Motor Spinal Nerve then travels to a Target Muscle in the PNS.

SPINAL CORD TRACT NAMES

- Indicate the Tract's Place of Origin and Destination.

ASCENDING SENSORY SPINAL CORD TRACTS

- DORSAL COLUMNS
- LATERAL SPINOTHALAMIC
- ANTERIOR SPINOTHALAMIC

- POSTERIOR SPINOCEREBELLAR
- ANTERIOR SPINOCEREBELLAR
- CUNEOCEREBELLAR
- ROSTRAL SPINOCEREBELLAR

CORTICALLY ORIGINATED DESCENDING MOTOR TRACTS

- LATERAL CORTICOSPINAL
- ANTERIOR CORTICOSPINAL
- CORTICOBULBAR

MIXED PATHWAYS ORIGINATING FROM THE BRAINSTEM

- MEDIAL LONGITUDINAL FASCICULUS

DESCENDING MOTOR TRACTS ORIGINATING IN THE BRAINSTEM

- VESTIBULOSPINAL
- RUBROSPINAL
- MEDULLARY RETICULOSPINAL
- PONTINE RETICULOSPINAL

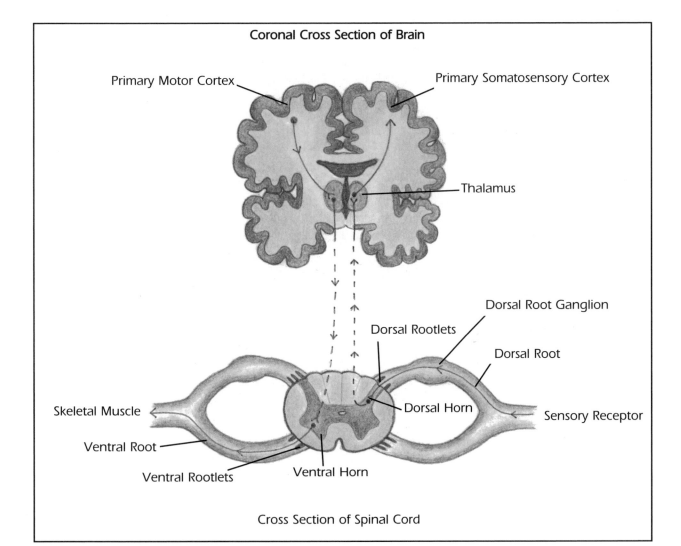

Coronal Cross Section of Brain

Primary Motor Cortex

Primary Somatosensory Cortex

Thalamus

Dorsal Root Ganglion

Dorsal Rootlets

Dorsal Root

Skeletal Muscle

Dorsal Horn

Sensory Receptor

Ventral Root

Ventral Rootlets

Ventral Horn

Cross Section of Spinal Cord

Ascending Sensory Spinal Cord Tracts

Dorsal Columns (Also Called Medial Lemniscus or Posterior Columns)
- BELIEVED TO BE A NEWER TRACT PHYLOGENETICALLY.

Function
- Carries Conscious Sensory Information to the Cortex regarding:
  - Discriminative Touch
  - Pressure
  - Vibration
  - Proprioception
  - Kinesthesia

Origin
- Nucleus Gracilus and Nucleus Cuneatus (in the Dorsal Horn).

Decussation (Where the Tracts Cross)
- Caudal Medulla Level.

Destination
- Postcentral Gyrus (Primary Somatosensory Area, SS1).

Pathway
- Skin Receptors in the PNS send Sensory Information along the Peripheral Spinal Nerves.
- The Sensory Spinal Nerves travel through the Dorsal Root and Rootlets (still in the PNS).
- The Spinal Nerves then Synapse on an Interneuron in the Dorsal Horn of the SC (in the CNS).
- The Interneuron then Synapses on the Cell Bodies of the Dorsal Column Tract in the Nucleus Gracilus and Cuneatus (in the Dorsal Horn) and the Tract begins to Ascend.
- The Dorsal Column Tract then Decussates across the Midline in the Caudal Medulla.
- The Tract continues to Ascend in the Dorsal Funiculus.
- The Tract Ascends through the SC, Brainstem, Thalamus, Internal Capsule, and synapses in the Postcentral Gyrus.

Lesions
- COMPLETE SEVERANCE OF THE SC:
  - Bilateral Loss of Sensation (Discriminitive Touch) Below the Severed Cord Level
- A HEMI-LESION (on one side of the SC) BELOW THE DECUSSATION:
  - Causes Ipsilateral Loss
  - Example: If the Right SC is severed at T1, causes Right Sided Sensory Loss
- A HEMI-LESION IN THE BRAINSTEM (above the Medulla Level):
  - Causes Contralateral Sensory Loss.
- A LESION IN THE CORTEX:
  - Causes Contralateral Sensory Loss.

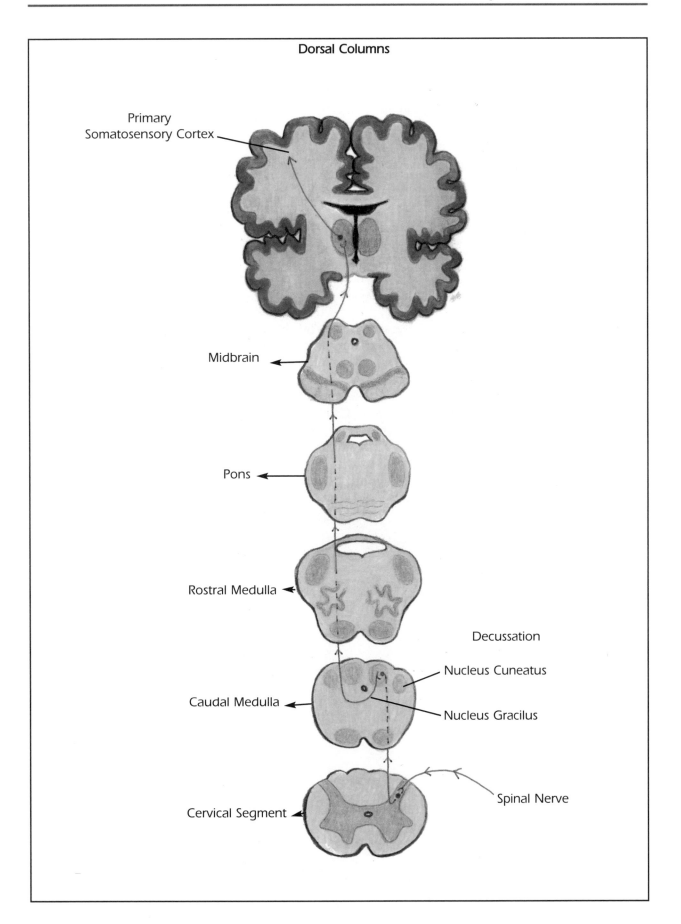

Dorsal Columns

Primary
Somatosensory Cortex

Midbrain

Pons

Rostral Medulla

Decussation

Nucleus Cuneatus

Caudal Medulla

Nucleus Gracilus

Spinal Nerve

Cervical Segment

Lateral Spinothalamic Tract
JOINS THE DORSAL COLUMN AT THE LEVEL OF THE PONS.

Function
- Carries Conscious Sensory Information to the Cortex regarding:
  - Pain and Temperature

Origin
- Nucleus Proprius in the Dorsal Horn.

Decussation
- Spinal Cord Level—Crosses as Soon as the Spinal Nerve enters the Cord and Synapses on the Tract.

Destination
- Postcentral Gyrus (Primary Somatosensory Area, SS1).

Pathway
- Skin Receptors in the PNS send Sensory Information along Peripheral Sensory Spinal Nerves.
- The Sensory Spinal Nerves travel through the Dorsal Root and Rootlets (in the PNS), and enter the Dorsal Horn (in the CNS).
- Once in the Dorsal Horn, the Spinal Nerves synapse on an Interneuron that joins to the Cell Bodies of the Lateral Spinothalamic Tract. This occurs in the Nucleus Proprius of the Dorsal Horn.
- The Lateral Spinothalamic Tract Crosses the Midline as soon as it Synapses in the Dorsal Horn. The Tract then travels to the Anterior White Funiculus.
- From the Anterior White Funiculus, the Tract travels to the Lateral White Funiculus and Ascends up the SC.
- The Tract Joins the Dorsal Column in the Pons and Continues to Ascend to the Thalamus, Internal Capsule, and finally synapses in the Postcentral Gyrus in the Cortex.

Lesions
- COMPLETE SEVERANCE OF THE SC:
  - Bilateral Loss of Sensation (Pain and Temperature) Below the Severed Cord Level
- A HEMI-LESION (on one side of the cord) IN THE SC:
  - At the Lesion Level: Bilateral Sensory Loss.
  - Below the Lesion Level: Contralateral Sensory Loss.
- A HEMI-LESION (on one side) OF THE BRAINSTEM:
  - Contralateral Sensory Loss.
- A UNILATERAL LESION IN THE POSTCENTRAL GYRUS:
  - Contralateral Sensory Loss.

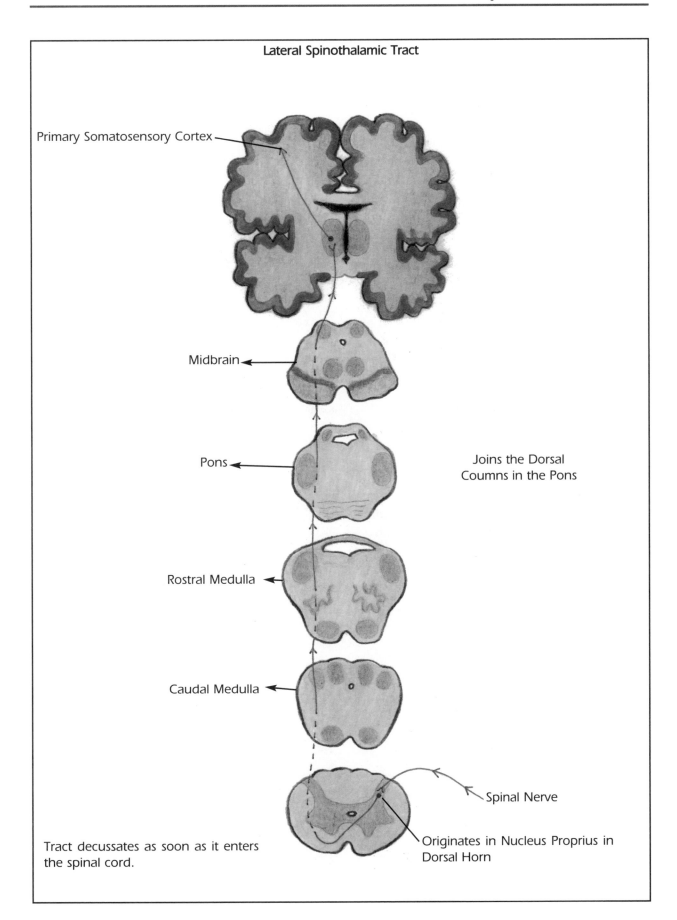

Lateral Spinothalamic Tract

Primary Somatosensory Cortex

Midbrain

Pons

Joins the Dorsal
Coumns in the Pons

Rostral Medulla

Caudal Medulla

Spinal Nerve

Originates in Nucleus Proprius in
Dorsal Horn

Tract decussates as soon as it enters
the spinal cord.

Anterior Spinothalamic Tract

Function

- Carries Conscious Sensory Information to the Cortex regarding:
  - Crude Touch, Light Touch
  - Some Theorists believe that this Tract may carry Pain and Temperature if the Lateral Spinothalamic is damaged

Origin

- Nucleus Proprius in the Dorsal Horn.

Decussation

- SC Level (the Tract crosses the midline as soon as spinal nerves have synapsed on its Cell Bodies in the Dorsal Horn).

Destination

- Postcentral Gyrus (Primary Somatosensory Area, SS1).

Pathway

- Skin Receptors in the PNS send Sensory Messages along the Ascending Sensory Spinal Nerves to the Dorsal Root and Rootlets.
- The Sensory Spinal Nerves Synapse on an Interneuron in the Dorsal Horn (specifically in the Nucleus Proprius).
- The Interneuron then Synapses on the Cell Bodies of the Ascending Spinothalamic Tract.
- At this time the Tract crosses the midline of the SC and travels to the Ventral White Funiculus and begins to Ascend.
- As is Ascends, the Tract travels through the Posterolateral Funiculus and eventually joins the Lateral Spinothalamic Tract in the Medulla.
- The Tract Ascends through the Brainstem, Thalamus, Internal Capsule, to the Primary Somatosensory Cortex (SS1).

Lesions

- COMPLETE SEVERANCE OF THE SC:
  - Bilateral Loss of Sensation (Crude and Light Touch) Below the Severed Cord Level
- A HEMI-LESION (on one side) OF THE SPINAL CORD:
  - At the Lesion Level: Bilateral Sensory Loss
  - Below the Lesion Level: Contralateral Sensory Loss.
- A HEMI-LESION (on one side) OF THE BRAINSTEM:
  - Contralateral Sensory Loss
- A UNILATERAL LESION IN THE POSTCENTRAL GYRUS:
  - Contralateral Sensory Loss

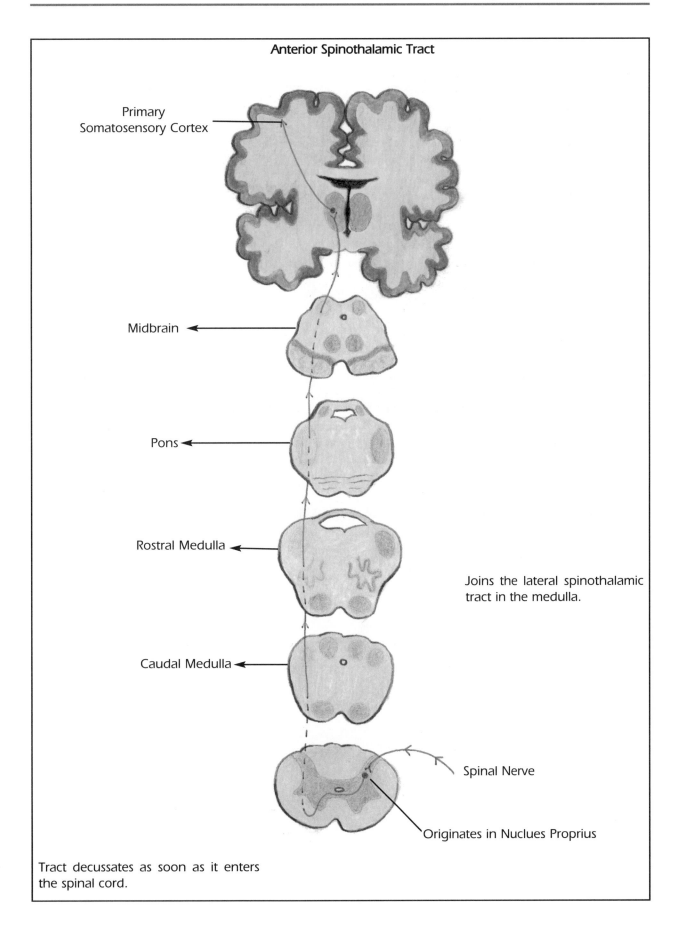

Anterior Spinothalamic Tract

Primary
Somatosensory Cortex

Midbrain

Pons

Rostral Medulla

Joins the lateral spinothalamic
tract in the medulla.

Caudal Medulla

Spinal Nerve

Originates in Nuclues Proprius

Tract decussates as soon as it enters
the spinal cord.

Posterior Spinocerebellar Tract

Function

- Carries Unconscious Sensory Information from the Lower Extremities to the Cerebellum regarding:
  - Proprioception
    - Pressure and Tension of Skeletal Muscles
    - Coordination of Motoric Movement of Individual Muscles
- Carries Sensory Information from the Muscle Spindles (MS), Golgi Tendon Organs (GTO), and the Joint Receptors in the PNS
- This information never reaches the Cortex for Conscious Detection and Interpretation

Origin

- Clark's Column in the Dorsal Horn in the SC Levels T6 and Below (because this tract serves the Lower Extremities).
- Clark's Column contains the Lower Extremity Proprioceptive and Kinesthetic Inputs.

Decussation

- NONE. This is an Ipsilateral Tract that does NOT Cross.

Destination

- Cerebellum.

Pathway

- The MSs, GTOs, and Joint Receptors in the PNS send Proprioceptive Information along the Ascending Sensory Spinal Nerves to the Dorsal Root and Rootlets.
- The Sensory Spinal Nerves then synapse on an Interneuron in the Dorsal Horn (specifically in Clark's Column).
- The Interneuron joins to the Cell Bodies of the Posterior Spinocerebellar Tract in the Dorsal Horn and the Tract begins to Ascend up the Posterolateral Funiculus on the Ipsilateral Side.
- The Tract Ascends through the Medulla to the Inferior Cerebellar Peduncle.
- From the Inferior Cerebellar Peduncle the Tract synapses in the Cerebellum.

Lesions

- All Lesions are Ipsilateral because the Tract does not Decussate.

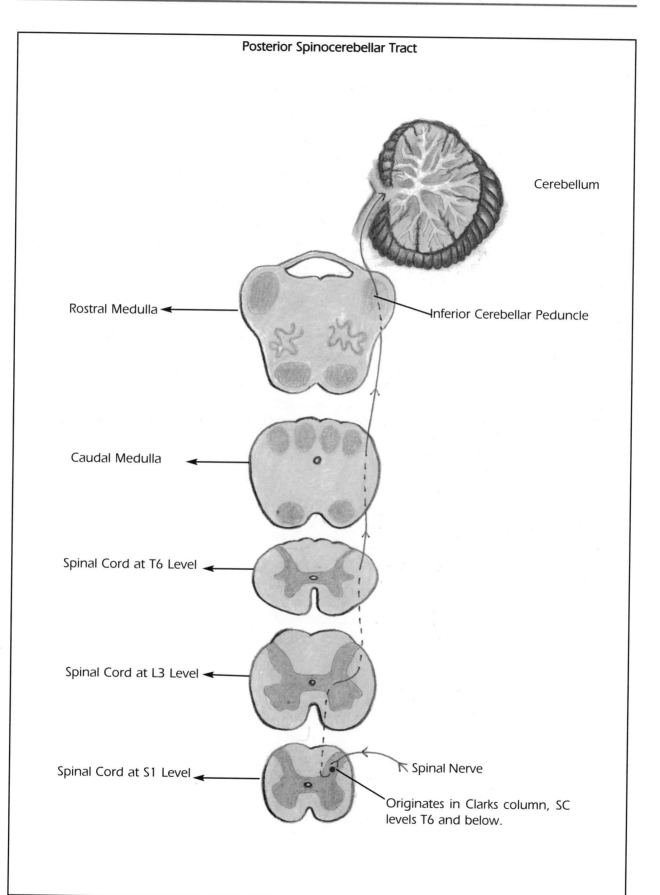

Posterior Spinocerebellar Tract

Cerebellum

Rostral Medulla

Inferior Cerebellar Peduncle

Caudal Medulla

Spinal Cord at T6 Level

Spinal Cord at L3 Level

Spinal Cord at S1 Level

Spinal Nerve

Originates in Clarks column, SC levels T6 and below.

Anterior Spinocerebellar Tract

Function

- Carries Unconscious Sensory Information from the Lower Extremities to the Cerebellum regarding:
  - Proprioception
    - Pressure and Tension of Skeletal Muscles
    - Coordination of Posture and Movement of Limbs (not individual muscles as in the Posterior Spinocerebellar Tract)
- Carries Sensory Information from the MSs, GTOs, and Joint Receptors, to the Cerebellum.

Origin

- Nucleus Proprius in the Dorsal Horn of the Lumbar Sections (because the Tract serves the Lower Extremities).

Decussation

- Spinal Cord Level in the Lumbar Sections. The Tract Decussates as soon as the Spinal Nerves Synapse on the Cell Bodies of the Anterior Spinocerebellar Tract.

Destination

- Cerebellum.

Pathway

- The MSs, GTOs, and Joint Receptors in the PNS send Proprioceptive Information along the Ascending Sensory Spinal Nerves to the Dorsal Root and Rootlets.
- The Sensory Spinal Nerves synapse on an Interneuron in the Dorsal Horn (specifically the Nucleus Proprius).
- The Interneuron then synapses on the Cell Bodies of the Anterior Spinocerebellar Tract in the Dorsal Horn and the Tract then Decussates across the midline to the Anterior White Funiculus.
- The Tract begins to Ascend in the Lateral White Funiculus and travels to the Superior Cerebellar Peduncle in the Pons.
- From the Superior Cerebellar Peduncle the Tract synapses in the Cerebellum.

Lesions

- COMPLETE SEVERANCE OF THE SC:
  - Bilateral Loss of proprioception from the LEs
- A HEMI-LESION (on one side) OF THE SPINAL CORD:
  - At the Lesion Level: Bilateral Proprioceptive Loss
  - Below the Lesion Level: Contralateral Proprioceptive Loss
- A LESION IN THE SUPERIOR CEREBELLAR PEDUNCLE:
  - Contralateral Proprioceptive Loss
- A LESION IN ONE HEMISPHERE OF THE CEREBELLUM:
  - Contralateral Proprioceptive Loss

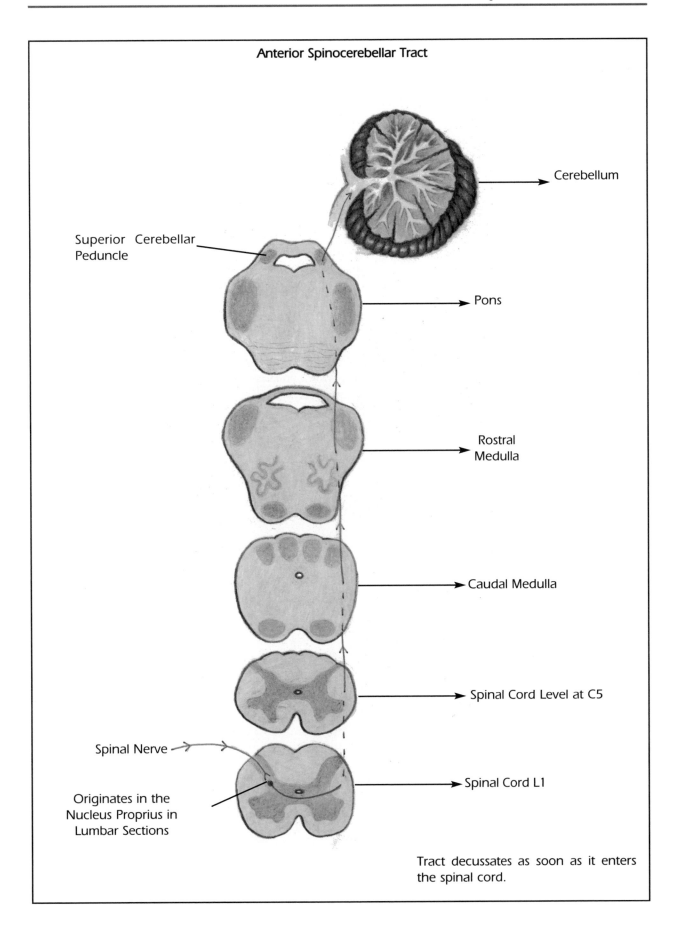

Anterior Spinocerebellar Tract

Cerebellum

Superior Cerebellar Peduncle

Pons

Rostral Medulla

Caudal Medulla

Spinal Cord Level at C5

Spinal Nerve

Spinal Cord L1

Originates in the Nucleus Proprius in Lumbar Sections

Tract decussates as soon as it enters the spinal cord.

Cuneocerebellar Tract

Function

- Carries Unconscious Sensory Information from the Trunk and Upper Extremities to the Cerebellum regarding:
  - Proprioception
    - Pressure and Tension of Skeletal Muscles
    - Coordination of Motoric Movement of Individual Muscles
- Carries Sensory Information from the MSs, GTOs, and the Joint Receptors to the Cerebellum.

Origin

- Nucleus Cuneatus in the Dorsal Horn of the Spinal Cord of T6 and Above (because the Tract serves the Upper Extremities).

Decussation

- NONE. This is an Ipsilateral Tract that does NOT Cross.

Destination

- Cerebellum.

Pathway

- The MSs, GTOs, and Joint Receptors in the PNS send Proprioceptive Information along the Ascending Sensory Spinal Nerves to the Dorsal Root and Rootlets of T6 and Above.
- The Sensory Spinal Nerves synapse on an Interneuron in the Dorsal Horn (specifically in the Nucleus Cuneatus).
- The Interneuron then synapses with the Cell Bodies of the Cuneocerebellar Tract in the Dorsal Horn.
- The Tract begins to Ascend in the Posterolateral White Funiculus and travels to the Inferior Cerebellar Peduncle in the Medulla.
- From the Inferior Cerebellar Peduncle, the Tract synapses in the Cerebellum.

Lesions

- All Lesions are Ipsilateral because the Tract does NOT Decussate.

Bilateral Representation of Unconscious Proprioception

- The below two sets of Tracts Provide Bilateral Representation of Unconscious Proprioceptive Information to the Cerebellum.
  - PROPRIOCEPTION FROM THE LOWER EXTREMITIES:
    - Posterior Spinocerebellar
    - Anterior Spinocerebellar
  - PROPRIOCEPTION FROM THE TRUNK AND UPPER EXTREMITIES:
    - Cuneocerebellar
    - Rostral Cerebellar

If one Tract of each set is lost, the other will still send Proprioceptive information to the Cerebellum. An individual will not lose Unconscious Proprioceptive Information from the Upper or Lower Extremities unless Both Tracts within a Set are Lost.

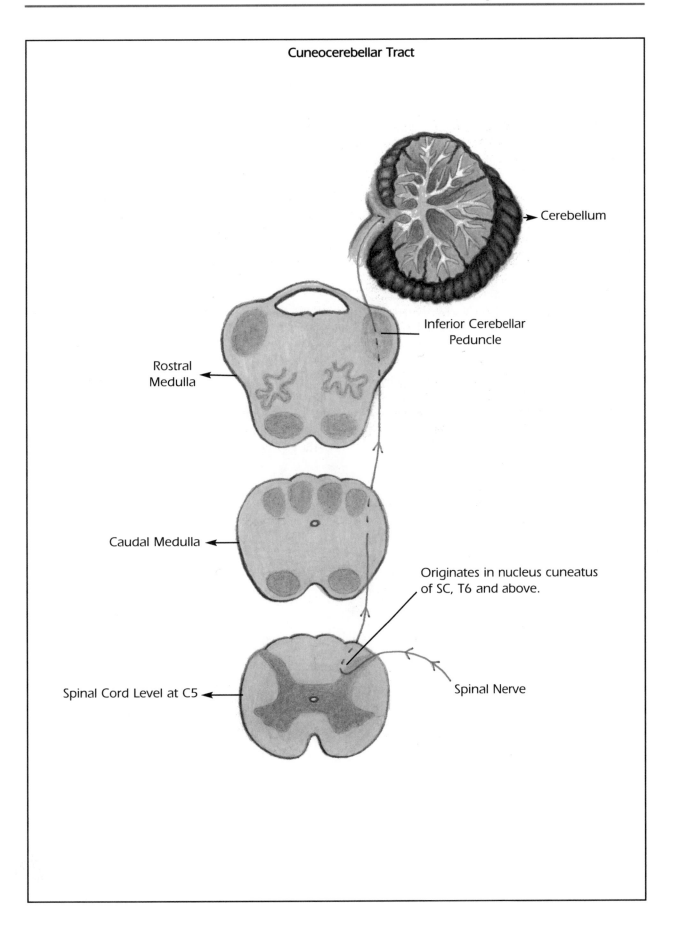

Cuneocerebellar Tract

Cerebellum

Inferior Cerebellar Peduncle

Rostral Medulla

Caudal Medulla

Originates in nucleus cuneatus of SC, T6 and above.

Spinal Nerve

Spinal Cord Level at C5

Rostral Spinocerebellar Tract

Function

- Carries Unconscious Sensory Information from the Trunk and Upper Extremities to the Cerebellum regarding:
  - Proprioception
    - Pressure and Tension of Skeletal Muscles
    - Coordination of Posture and Movements of Limbs
  - Carries Sensory Information from the MSs, GTOs, and Joint Receptors to the Cerebellum

Origin

- Ventrolateral Gray of the Spinal Cord in the Cervical Levels (because the Tract serves the Upper Extremities).

Decussation

- NONE. This is an Ipsilateral Tract that does NOT Cross.

Destination

- Cerebellum.

Pathway

- The MSs, GTOs, and Joint Receptors in the PNS send Proprioceptive Information along the Ascending Sensory Spinal Nerves to the Dorsal Root and Rootlets of the Cervical Levels.
- The Sensory Spinal Nerves enter the Dorsal Horn of the SC and Synapse in the Ventrolateral Gray on an Interneuron.
- The Interneuron then synapses on the Cell Bodies of the Rostral Spinocerebellar Tract and the Tract begins to Ascend up the Lateral White Funiculus.
- The Tract joins the Anterior Spinocerebellar Tract in the Lateral White Funiculus.
- The Tract then Ascends to either the Inferior Cerebellar Peduncle or the Superior Cerebellar Peduncle.
- From one of these Peduncles, the Tract then synapses in the Cerebellum.

Lesions

- All Lesions are Ipsilateral.

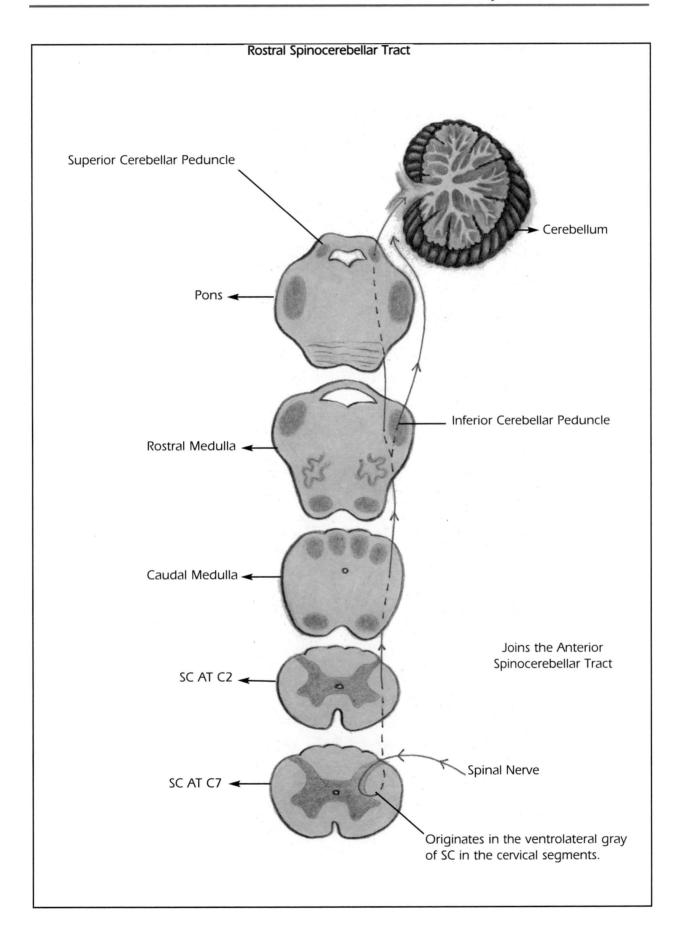

Rostral Spinocerebellar Tract

Superior Cerebellar Peduncle

Cerebellum

Pons

Inferior Cerebellar Peduncle

Rostral Medulla

Caudal Medulla

Joins the Anterior
Spinocerebellar Tract

SC AT C2

Spinal Nerve

SC AT C7

Originates in the ventrolateral gray
of SC in the cervical segments.

CORTICALLY ORIGINATED DESCENDING MOTOR TRACTS

Lateral and Anterior Corticospinal Tracts

Part of the Pyramidal System
- The Lateral and Anterior Corticospinal Tracts are Descending Motor Tracts that use the Pyramids of the Medulla.
- The Lateral Corticospinal Tract Decussates at the Pyramidal Decussation; the Anterior Corticospinal Tract remains Ipsilateral and descends through the Pyramids.
- These Tracts are Considered to be Upper Motor Neurons (UMNs) in the CNS.

Function
- Carry Conscious/Voluntary Motor Information from the Precentral Gyrus (Primary Motor Area, M1) up to, but not including the Ventral Horn.
- The Tracts then Synapse on Motor Spinal Nerves (in the Ventral Horn) that innervate Skeletal Muscles.

Origin
- Precentral Gyrus (Primary Motor Area, M1).

Decussation
- LATERAL CORTICOSPINAL TRACT:
  - Decussates in the Pyramidal Decussation of the Medulla.
- ANTERIOR CORTICOSPINAL TRACT:
  - Does NOT Decussate.

Destination
- Synapses on an Interneuron in the Ventral Horn.
- This Interneuron then Synapses on the Motor Neurons of the Motor Spinal Nerves that innervate Skeletal Muscles in the PNS.

Pathway
- The Cell Bodies of the Tract are Located in the Pre-Central Gyrus. They send Descending Motor Signals through the Internal Capsule, Thalamus, and Brainstem
- In the Brainstem (at the Medulla Level) the Tract separates into the Lateral and Anterior Corticospinal Tracts.
- The Lateral Corticospinal Tracts (90% of the Original Tract before Separation) Decussates in the Pyramidal Decussation in the Medulla.
- The Anterior Corticospinal Tract (10% of the Original Tract before Separation) remains Ipsilateral and continues to descend through the Pyramids.
- When the Tract is ready to exit the Cord, it synapses on an Interneuron in the Ventrolateral Gray of the SC.
- The Interneuron then synapses on a Motor Neuron (in the Ventral Horn) that innervates a Descending Motor Spinal Nerve.
- The Motor Spinal Nerve then exits through the Ventral Rootlets and Root and travels to a Target Skeletal Muscle in the PNS.

Lesions
- A LESION IN THE PRIMARY MOTOR AREA (M1) (in one hemisphere):
  - Results in Contralateral Loss of Voluntary Muscle Movement (because 90% of the Tract decussates).
  - Results in Spasticity of the Distal Musculature Below the Lesion Level (UMN Damage).
  - Results in Hyperactive Reflexes.
- A LESION IN THE INTERNAL CAPSULE:
  - Results in Contralateral Spastic Paralysis.
  - Hyperactive Reflexes.

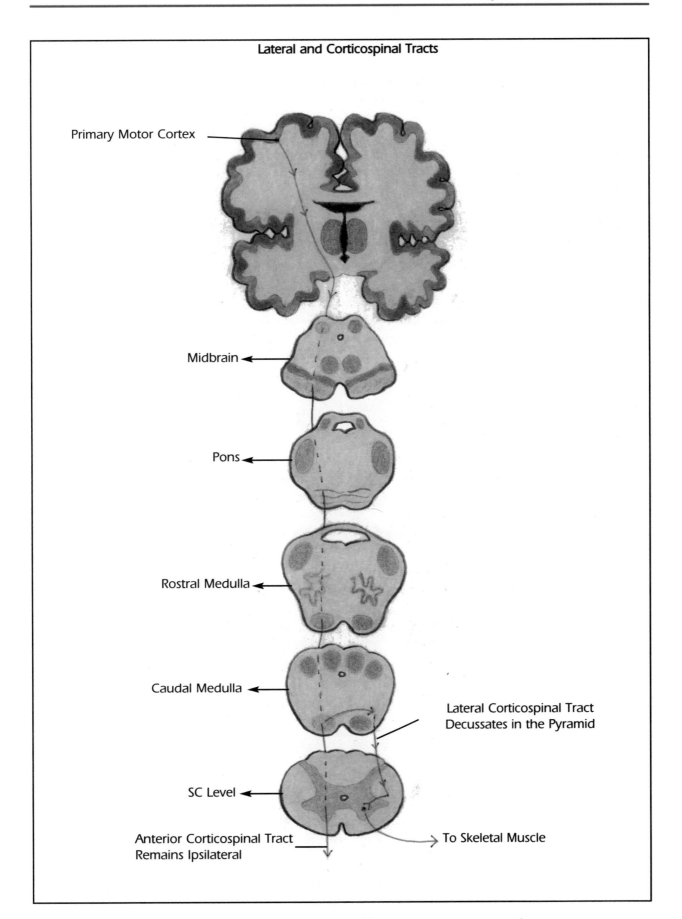

Lateral and Corticospinal Tracts

Primary Motor Cortex

Midbrain

Pons

Rostral Medulla

Caudal Medulla

Lateral Corticospinal Tract
Decussates in the Pyramid

SC Level

Anterior Corticospinal Tract
Remains Ipsilateral

To Skeletal Muscle

- A UNILATERAL LESION IN THE BRAINSTEM ABOVE THE DECUSSATION:
  - Contralateral Spastic Paralysis.
- A UNILATERAL LESION IN THE SPINAL CORD BELOW THE DECUSSATION:
  - Ipsilateral Loss of Voluntary Motor Control.
  - Spasticity in the Distal Musculature (Below the Lesion Level).
  - Flaccidity (At the Lesion Level).
  - Hyperactive Reflexes.
- A COMPLETE SEVERANCE OF THE SC BELOW THE DECUSSATION
  - Bilateral Loss of Voluntary Motor Control
  - Spasticity Below the Lesion Level
  - Flaccidity At the Lesion Level

Decorticate Rigidity

- Damage to the Corticospinal Tracts will Result in Decorticate Rigidity.
- Decorticate Rigidity Presents as:
  - Upper Extremities are in a Spastic Flexed Position
  - Lower Extremities are in a Spastic Extended Position
- Decorticate Rigidity occurs because the Corticospinal Tracts fire without Modification from the Cortex.

Corticobulbar Tract

Pathway

- The Corticobulbar Tract Descends from the Corticospinal Tract and projects to certain Cranial Nerve Nuclei that have a Motor Component:
  - In the Midbrain, the Corticobulbar Tract projects to Cranial Nerve Nuclei 3, 4, 6 (the Extraocular CNs).
  - In the Pons, the Corticobulbar Tract projects to Cranial Nerve Nuclei 5, 7 (Facial Muscles).
  - In the Medulla, the Corticobulbar Tract projects to Cranial Nerve Nuclei 9, 10, 11, 12 (Muscles for Swallowing, Eating, Speaking).

Function

- The Corticobulbar Tracts Control Cranial Nerve Lower Motor Neurons.

Lesions

- Lesions involve the above listed Cranial Nerve Nuclei.
- Lesions result in Flaccidity of the Anatomy innervated by the above Cranial Nerves.

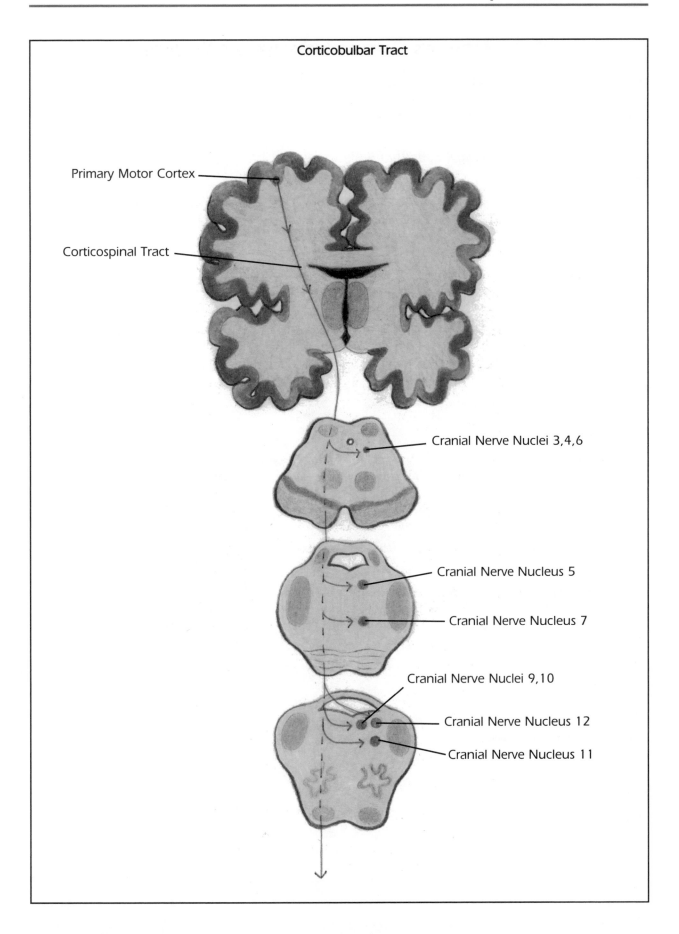

Corticobulbar Tract

Primary Motor Cortex

Corticospinal Tract

Cranial Nerve Nuclei 3,4,6

Cranial Nerve Nucleus 5

Cranial Nerve Nucleus 7

Cranial Nerve Nuclei 9,10

Cranial Nerve Nucleus 12

Cranial Nerve Nucleus 11

MIXED PATHWAYS ORIGINATING FROM THE BRAINSTEM

Medial Longitudinal Fasciculus

- A Fiber Bundle situated near the Midline of the Brainstem.
- Composed of Ascending and Descending Fibers.
- Part of the UMN System.

Descending Fibers of the Medial Longitudinal Fasciculus

Function

- The MLF exerts an Inhibitory Effect on the Motor Neurons of the Ventral Horn in the Cervical Spinal Cord.
- This Inhibitory Effect allows for the Coordination of Head and Neck Movements.

Origin

- Vestibular Nuclei in the Medulla.

Decussation

- NONE. This is an Ipsilateral Tract that does NOT Cross.

Destination

- From the Cervical Segments of the Spinal Cord, the Tract projects to Spinal Nerves in the PNS that innervate Head and Neck Musculature.

Pathway

- The Vestibular Nuclei in the Medulla (the Origin of the Tract) Receive Information from:
  - The Superior Colliculi in the Midbrain
  - The Oculomotor Nuclei in the Midbrain
  - The Pontine Reticular Formation
- The Tract then Descends through the Anterior White Funiculus of the SC.
- When the Tract is ready to Exit the Cord, it synapses on an Interneuron in the Ventral Horn of the Cervical Cord Segments.
- The Interneuron then synapses on Motor Neurons (in the Ventral Horn) of Descending Motor Spinal Nerves that Innervate Head and Neck Musculature in the Periphery.

Ascending Fibers of the Medial Longitudinal Fasciculus

Function

- The Ascending Fibers of the MLF are responsible for the Visual Tracking of a Moving Object through the Coordinated Movements of the Eyes, Head, and Neck.
- The Ascending Fibers of the MLF are responsible for the Refinement of Extraocular Eye Movements.

Origin

- Vestibular Nuclei of the Medulla.

Decussation

- NONE. This is an Ipsilateral Tract that does NOT Cross.

Destination

- Oculomotor Nerve Nuclei.
- Trochlear Nerve Nuclei.
- Abducens Nerve Nuclei.

Pathway

- The Tract begins in the Vestibular Nuclei in the Medulla and Terminates on the Cranial Nerve Nuclei 3, 4, 6.

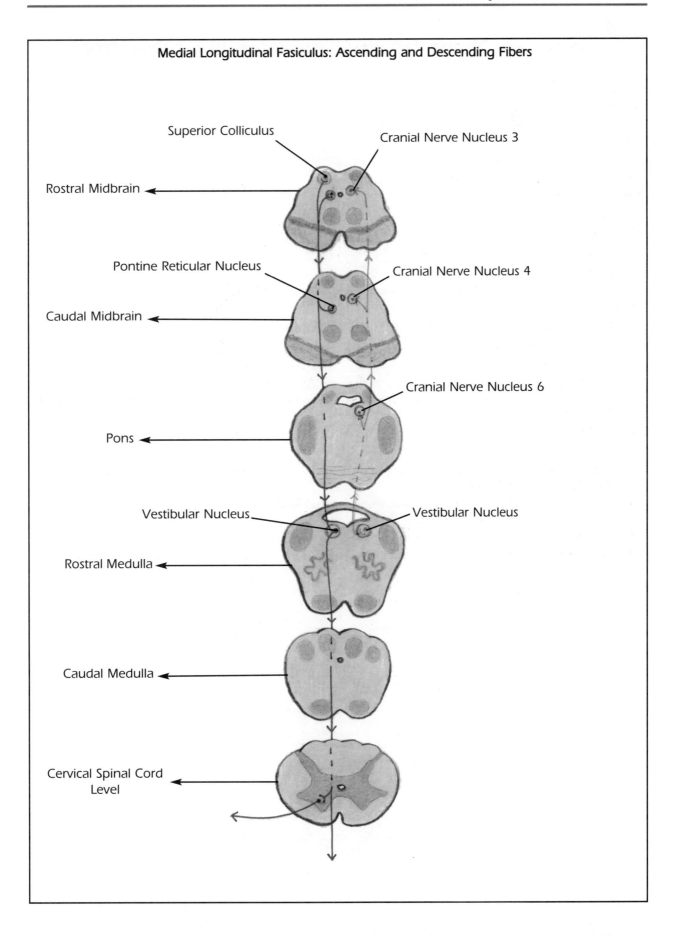

Medial Longitudinal Fasiculus: Ascending and Descending Fibers

Superior Colliculus

Cranial Nerve Nucleus 3

Rostral Midbrain

Pontine Reticular Nucleus

Cranial Nerve Nucleus 4

Caudal Midbrain

Cranial Nerve Nucleus 6

Pons

Vestibular Nucleus

Vestibular Nucleus

Rostral Medulla

Caudal Medulla

Cervical Spinal Cord
Level

Descending Motor Pathways Originating in the Brainstem

Vestibulospinal Tract

Extrapyramidal Tract

- The Vestibular Tract does NOT use the Pyramids and is thus considered to be an Extrapyramidal Tract.

Function

- Facilitation of Antigravity (Extensor) Muscles.
- Facilitation of Muscles responsible for Posture and Stance.

Origin

- The Vestibular Nuclei in the Medulla.

Decussation

- NONE. The Tract remains Ipsilateral.

Destination

- The Vestibulospinal Tract Innervates Extensor Muscle Groups in the PNS.

Pathway

- The Vestibular Nuclei in the Medulla (the Tract's Origin) Receive Messages from:
  - Vestibular Apparatus in the Inner Ear via the Vestibulocochlear Nerve
  - Cerebellum
- From the Vestibular Nuclei, the Tract then travels to the Anterolateral White Funiculus of the SC.
- When the Tract is ready to exit the Cord, it Synapses on an Interneuron that joins with a Motor Neuron in the Ventral Horn.
- The Motor Neuron synapses on a Descending Motor Spinal Nerve that innervates a Target Extensor Muscle.

Lesions

- Damage to any of the Tracts that Originate in the Brainstem—including the Vestibulospinal Tract—results in DECEREBRATE RIGIDITY.
- When Damage occurs to the Brainstem, the Vestibulospinal Tract fires without Modification from the Brainstem and Cortex, causing Decerebrate Rigidity.
- Decerebrate Rigidity involves Spastic Extension of both the Upper and Lower Extremities.
- The occurrence of Decerebrate Rigidity indicates a much poorer Prognosis than does Decorticate Rigidity.

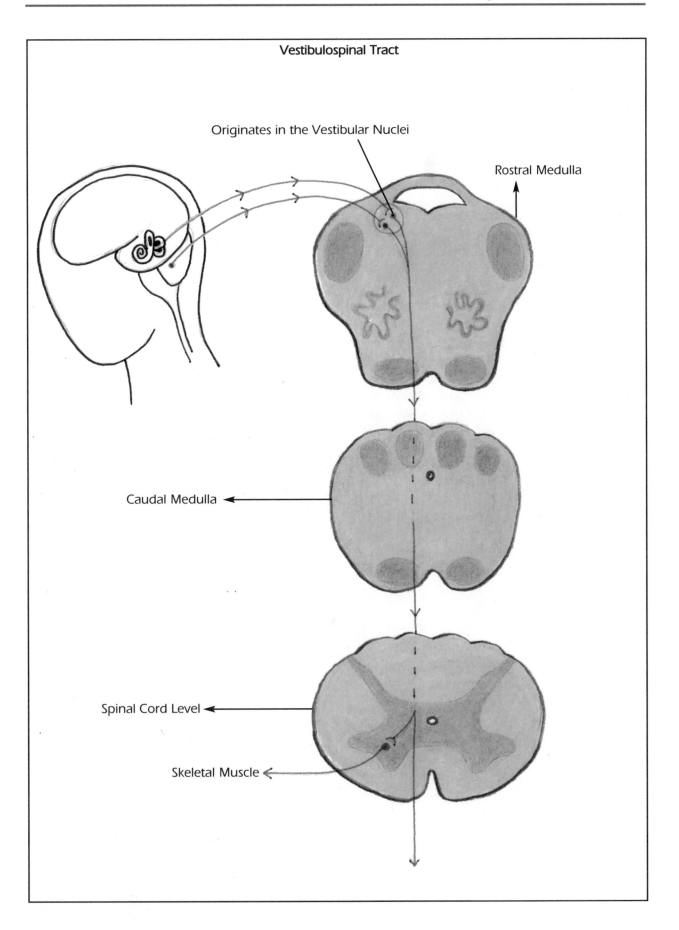

Vestibulospinal Tract

Originates in the Vestibular Nuclei

Rostral Medulla

Caudal Medulla

Spinal Cord Level

Skeletal Muscle

Rubrospinal Tract

Extrapyramidal Tract

- The Rubrospinal Tract does NOT use the Pyramids and is thus considered to be an Extrapyramidal Tract.

Function

- Facilitates the Antagonist of the Antigravity Muscles (Mostly in the Limbs)—Facilitation of Flexor Muscle Groups.
- The Activity of the Rubrospinal Tract is Modified by the Cerebellum and the Cortex.
- The Cerebellum and Cortex Modify the Activity of the Red Nucleus in the Midbrain—the Origin of the Rubrospinal Tract.

Origin

- Red Nucleus in the Midbrain.

Decussation

- Occurs at the Midbrain Level.

Destination

- The Rubrospinal Tract Innervates Flexor Muscles in the Limbs.

Pathway

- The Tract Begins in the Red Nucleus of the Midbrain.
- In the Midbrain, the Tract Crosses the Midline and Enters the Crus Cerebri.
- Once the Tract Decussates, it begins to Descend to the SC.
- The Rubrospinal Tract Travels very close to the Corticospinal Tract as it descends in the SC.
- When the Tract is ready to Exit the Cord, it synapses with an Interneuron in the Ventral Horn.
- The Interneuron joins with a Descending Motor Spinal Nerve that innervates Flexor Muscles in the PNS.

Lesions

- A Lesion to the Rubrospinal Tract results in DECEREBRATE RIGIDITY (involves Spastic Extension of the Upper and Lower Extremities).
- If the Rubrospinal Tract is lost, the Antagonist of the Antigravity Muscles are gone—the Flexors are Lost.
- The Upper and Lower Extremities present with Increased Extensor Tone.

Rubrospinal Tract

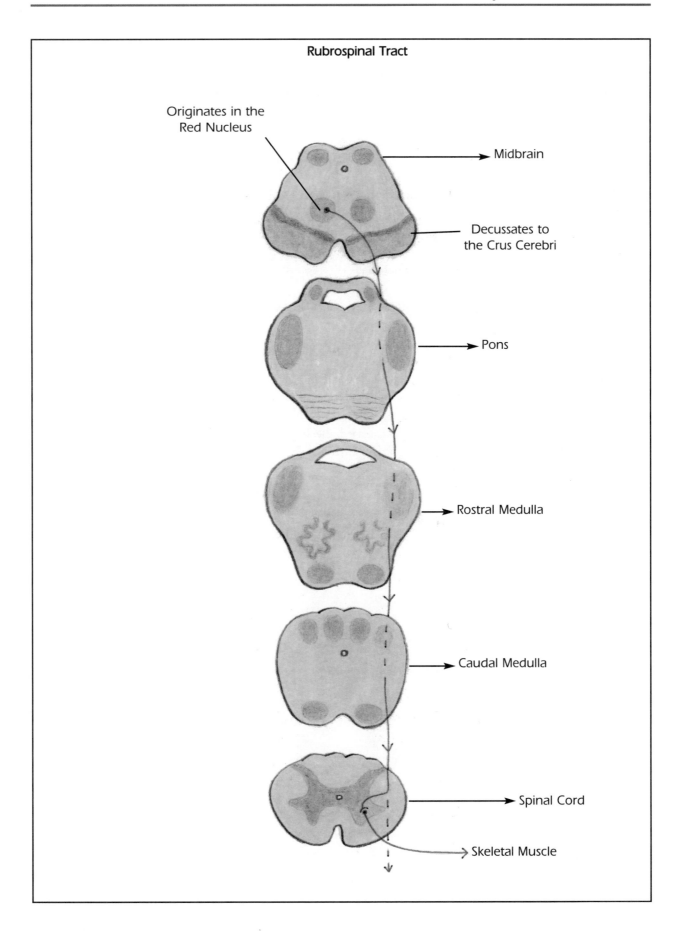

Originates in the
Red Nucleus

Midbrain

Decussates to
the Crus Cerebri

Pons

Rostral Medulla

Caudal Medulla

Spinal Cord

Skeletal Muscle

Medullary Reticulospinal Tract

Extrapyramidal Tract

- The Medullary Reticulospinal Tract does NOT use the Pyramids and is thus considered to be an Extrapyramidal Tract.

Function

- Inhibits Antigravity Muscles—inhibits Extensor Tone.
- This Tract also Depresses Cardiovascular Responses (Blood Pressure, Heart Rate) and the Inspiratory Phase of Respiration.
- The Tract receives substantial Modification from the Corticospinal Tracts.

Origin

- Medullary Reticular Formation (in the Medulla).

Decussation

- Most of the Fibers of this Tract Remain Uncrossed.

Destination

- The Medullary Reticulospinal Tract Modifies Antigravity or Extensor Muscles in the PNS.

Pathway

- The Tract Begins in the Medullary Reticular Formation in the Medulla.
- It Descends through the Anterior White Funiculus.
- When the Tract is ready to Exit the Cord, it synapses on an Interneuron in the Ventral Horn.
- The Interneuron then synapses on a Motor Spinal Nerve that travels to an Extensor Muscle in the PNS.

Lesions

- When the Medullary Reticulospinal Tract is damaged, the Tract that inhibits Extensor Tone is Lost.
- This causes an Increase in Extensor Tone, or DECEREBRATE RIGIDITY (Spastic Extension of the Upper and Lower Extremities).

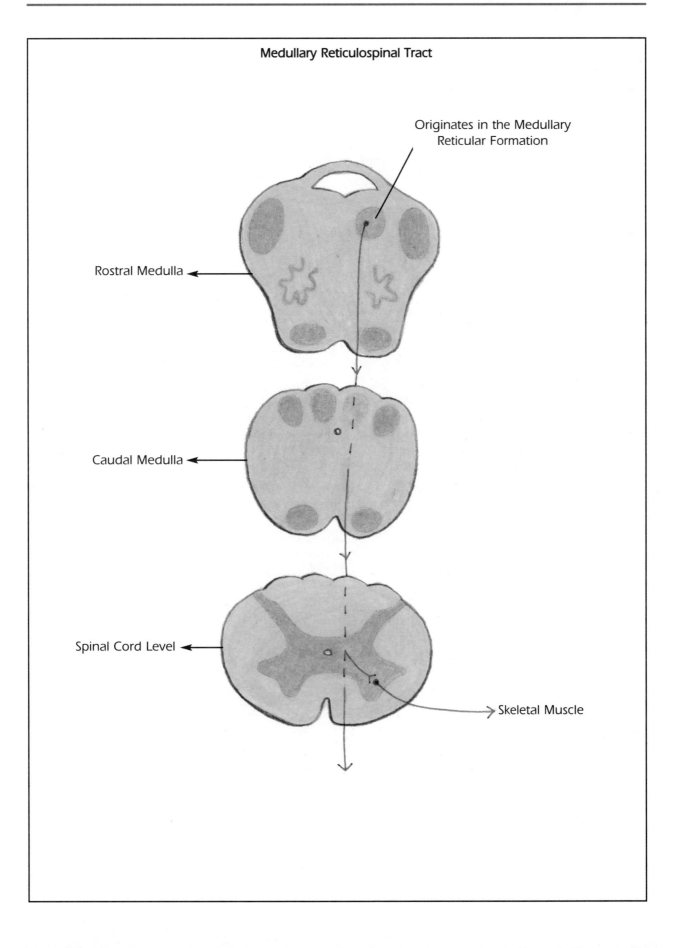

Medullary Reticulospinal Tract

Originates in the Medullary
Reticular Formation

Rostral Medulla

Caudal Medulla

Spinal Cord Level

Skeletal Muscle

Pontine Reticulospinal Tract

Extrapyramidal Tract

- The Pontine Reticulospinal Tract does NOT use the Pyramids and is thus considered to be an Extrapyramidal Tract.

Function

- Facilitates Antigravity Muscles—facilitates Extensor Tone.
- This Tract Receives substantial Modification from the Corticospinal Tracts.

Origin

- Pontine Reticular Formation (in the Pons).

Decussation

- NONE. This is an Ipsilateral Tract.

Destination

- The Pontine Reticulospinal Tract Innervates Extensor Muscles.

Pathway

- The Tract begins in the Pontine Reticular Formation.
- It Descends through the Anterior White Funiculus of the SC.
- When the Tract is ready to Exit the Cord, it synapses on an Interneuron in the Ventral Horn.
- The Interneuron then synapses with a Motor Spinal Nerve that innervates an Extensor Muscle in the PNS.

Lesions

- Damage to this Tract results in DECEREBRATE RIGIDITY (Spastic Extension of the Upper and Lower Extremities).
- Usually, damage to this Tract is less important than damage to the Tracts and neural Regions Surrounding it.
- If Cortical or Brainstem Damage has occurred, then nothing modifies this Tract's firing rate. Consequently, the Tract fires without Higher Center Modification. This leads to Increased Extensor Tone.

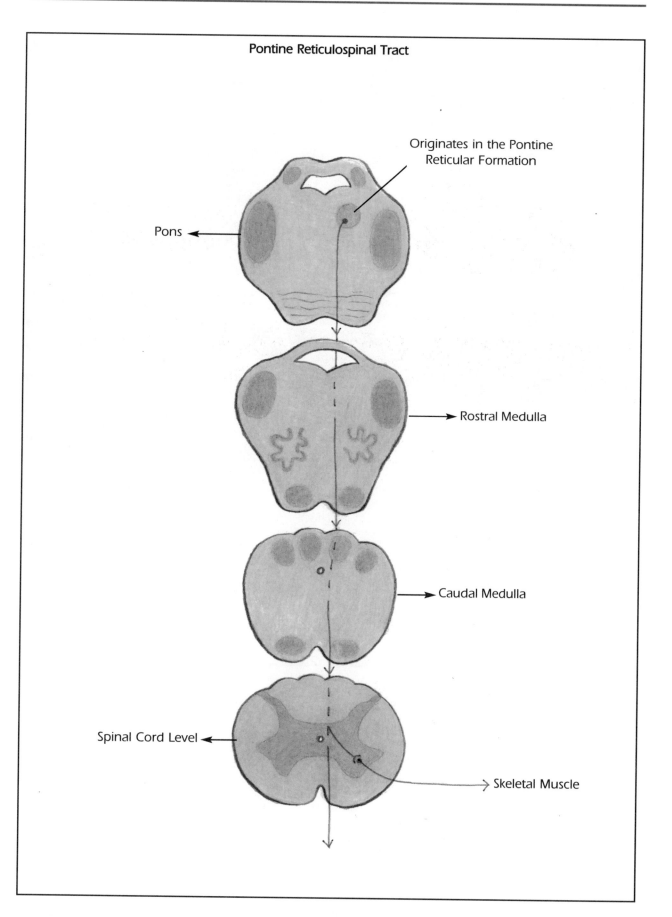

Pontine Reticulospinal Tract

Originates in the Pontine Reticular Formation

Pons

Rostral Medulla

Caudal Medulla

Spinal Cord Level

Skeletal Muscle

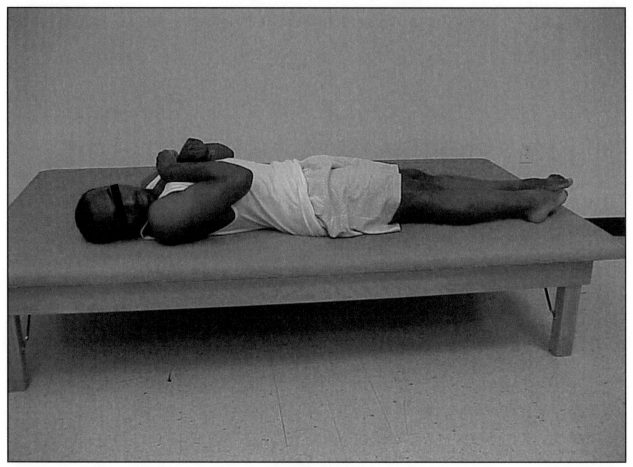

Decorticate Rigidity. Results from a lesion to the corticospinal tracts. The Upper Extremities are in spastic flexion: Head and Neck are flexed; Scapulae are elevated and retracted; Shoulders are internally rotated and adducted; Forearms are pronated; Elbows, Wrists, and Fingers are flexed; Wrists are ulnarly deviated. The Lower Extremities are in spastic extension: hips are rotated and adducted; Feet are plantar flexed and inverted.

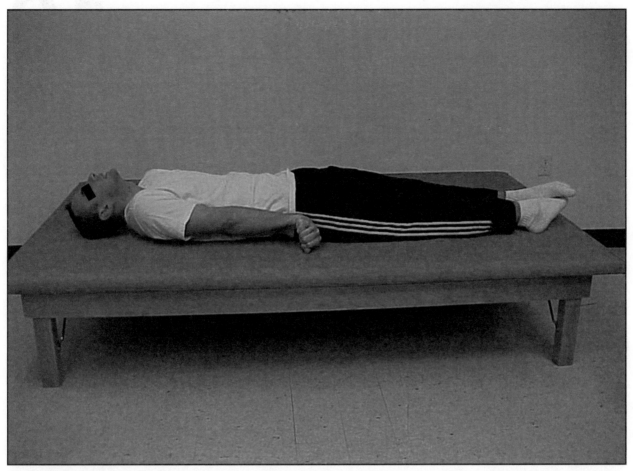

Decerebrate Rigidity. Results from a lesion to the extrapyramidal spinal cord tracts. Both the Upper and Lower Extremities are in spastic extesion. Head and Neck are in hyperextension; Jaws are clenched; Scapulae are elevated and retracted; Shoulders are internally rotated and adducted; Elbows are extended; Forearms are pronated; Wrists are flexed and ulnarly deviated; Fingers are flexed and adducted; Hips and Knees are extended; Hips are internally rotated and adducted; Feet are plantar flexed and inverted.

Spinal Cord Injury and Disease

COMMON CAUSES OF SPINAL CORD INJURY/DISEASE

Transection
- A Transection is a Complete Severance of the Cord.
- Involves interruption of All Sensory and Motor Information At and Below the Lesion Level.
- Transections can result from Traumatic Injury including Auto Accidents, Knife Wounds, Gun Shot Wounds, Diving Accidents.

Compression
- Compression involves Impingement of the Cord.
- Symptoms depend on the Severity of the Injury.
- Compressions can result from Tumor, Vertebral Degenerative Joint Disease.

Infection
- Infection may compromise the integrity of the Cord.
- An example is Polio: Involves damage to the Cell Bodies in the Ventral Horn causing Lower Motor Neuron (LMN) Loss.

Degenerative Disorders
- Degenerative Diseases can damage the Motor Spinal Cord Tracts.
- One example is Amyotrophic Lateral Sclerosis (ALS): Results in Bilateral Degeneration of the Ventral Horn and Pyramidal Tracts.
- Involves both Upper (UMN) and Lower Motor Neuron Damage.

UPPER MOTOR NEURON VS LOWER MOTOR NEURON LESIONS

Definition of Upper Motor Neuron
- A Motor Neuron that carries Motor Information from the Cortex or Subcortical Regions to:
- The Cranial Nerve Nuclei in the Brainstem. The Cranial Nerve Nuclei are considered to be part of the UMN System. The Cranial Nerve Fibers that travel to Target Muscles are considered to be within the LMN System.
- Interneurons that Synapse with Motor Cell Bodies in the Ventral Horn. An UMN travels up to but does not enter the Ventral Horn. The Ventral Horn is considered to be part of the LMN System.

Definition of Lower Motor Neuron
- A Motor Neuron that carries information from the Motor Cell Bodies in the Ventral Horn to Skeletal Muscles.
- LMNs include the Cranial Nerves and the Cauda Equina (below L1 – L2 vertebrae). Thus, Cord Injuries at the L1 Vertebra area and Lower are considered LMN Injuries. Cord Injuries higher than L1 Vertebra are considered UMN Injuries.

> ## Five Most Important Tracts to Clinically Evaluate
>
> 1. LATERAL CORTICOSPINAL TRACTS: for Voluntary Motor Control on the Contralateral Side.
> 2. DORSAL COLUMNS: for Conscious Discriminative Touch, Pressure, Vibration, Proprioception on Contralateral Side.
> 3. LATERAL SPINOTHALAMIC TRACTs: for Conscious Pain and Temperature on Contralateral Side.
> 4. SPINOCEREBELLAR TRACTs: for Unconscious Proprioception.
> 5. VESTIBULOSPINAL TRACTS: for Facilitation of Extensor Tone. Important to assess in Traumatic Brain Injury.

SIGNS AND SYMPTOMS OF UPPER AND LOWER MOTOR NEURON LESIONS

Upper Motor Neuron Lesion Signs

Below the Lesion Level
- SPASTICITY.
  - An Increase in Muscle Tone with an Associated Inability to Voluntarily Control the Muscle.
  - Difficulty Actively and Passively Moving the Muscles on One Side of the Joint—but NOT Both Sides.
  - Either the FLEXORS or the EXTENSORS are Spastic, but not both. If Both the FLEXORS and the EXTENSORS display increased tone, RIGIDITY is occurring.
- HYPERACTIVE REFLEXES.
- CLONUS.
  - A Sustained Series of Rhythmic Jerks in a Muscle.
  - Usually caused by a Quick Stretch of the Spastic Muscle Group.

At the Lesion Level
- FLACCIDITY: Loss of Muscle Tone.

Lower Motor Neuron Lesion Signs
- FLACCIDITY.
- HYPOREFLEXIA.
- WITHIN A FEW WEEKS OF LMN INJURY, MUSCLES BEGIN TO ATROPHY.
  - Muscles Undergoing the Early Stages of Atrophy May Display:
    - FIBRILLATIONS: Fine Twitchings of Single Muscle Fibers that usually cannot be detected on Clinical Exam but can be identified on an Electromyogram.
    - FASCICULATIONS: Brief Contractions of Motor Units; can be Observed in Skeletal Muscle and detected on Clinical Exam.

EMERGENCE OF PRIMITIVE REFLEXES AFTER UPPER MOTOR NEURON DAMAGE

- Primitive Reflexes—which had been integrated by the Central Nervous System (CNS) in Infancy or Early Childhood—Often Re-Emerge as a result of Serious Neurologic Damage.
- For example, Damage to the Corticospinal Tracts often results in the Re-Emergence of the BABINSKI SIGN (an Extensor Plantar Reflex).
  - An Abnormal Babinski Sign—after Corticospinal Damage—can be Elicited by Stroking the Outer Border of the Plantar Surface of the Foot.
  - If the Babinski Sign is Present, Extension of the First Toe Occurs and is sometimes Accompanied by Fanning of the other Toes.
- HOFFMAN'S SIGN can also Indicate Serious Corticospinal Tract Damage.
  - Hoffman's Sign can be elicited by Flicking the Nail of the individual's third Finger.
  - When Hoffman's Sign is Present, there is a Prompt Adduction of the Thumb and Flexion of the Index Finger.

The Following Primitive Reflexes May Emerge After UMN Damage

Spinal Cord Level Reflexes (Emerge as a result of spinal cord motor tract damage)
- Flexor Withdrawal
- Extensor Thrust
- Crossed Extension

Brainstem Level Reflexes (Emerge as a result of Brainstem Damage)
- Asymmetrical Tonic Neck (ATNR)
- Symmetric Tonic Neck (STNR)
- Tonic Labyrinthine Reflex (TLR)
- Positive Support Reaction
- Associated Reactions (overflow of activity into the opposite limb; results from an inability to selectively Inhibit the interneurons that synapse with the motor cell bodies of the opposite limb).

The Following Reflexes May Be Lost After UMN Damage

Midbrain Level Reflexes
- Righting Reactions

Basal Ganglial Level Reflexes
- Protective Extension
- Equilibrium Reactions

Cortical Level Reflexes
- Optic Righting

SPINAL CORD DISEASE

DORSAL COLUMN DISEASE (Also called Posterior Cord Syndrome or Tabes Dorsales

Etiology
- Seen in patients with NEUROSYPHILLIS.

Pathology
- Dorsal columns are lost bilaterally.

Symptomatology
- Causes a bilateral loss of:
  - Tactile Discrimination
  - Vibration
  - Pressure
  - Proprioception (often accompanied by ataxia)

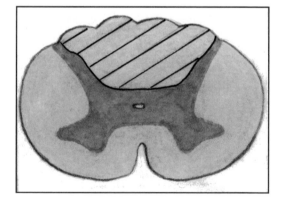

BROWN-SEQUARD SYNDROME

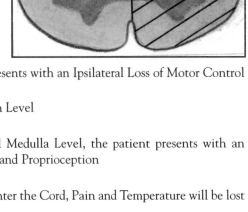

Etiology

- Multiple Sclerosis.
- Stab Wound.
- Tumor.

Pathology

- Brown-Sequard Syndrome is a Spinal Cord Hemisection.

Symptomatology

- The Lateral Corticospinal Tract is Lost Ipsilaterally.
    - Because the Injury Occurs in the Spinal Cord, the patient presents with an Ipsilateral Loss of Motor Control and Spasticity BELOW the Lesion Level
    - The patient presents with Ipsilateral Flaccidity AT the Lesion Level
- The Dorsal Column is Lost Ipsilaterally.
    - Because the Decussation of this Tract occurs at the Caudal Medulla Level, the patient presents with an Ipsilateral Loss of Discriminative Touch, Pressure, Vibration, and Proprioception
- The Spinothalamic Tract is Lost Contralaterally.
    - Because the Spinothalamic Tracts Decussate as soon as they enter the Cord, Pain and Temperature will be lost on the Contralateral Side (Below the lesion level)
    - At the Lesion Level, the patient experiences Bilateral Loss of Pain and Temperature

ANTERIOR CORD SYNDROME

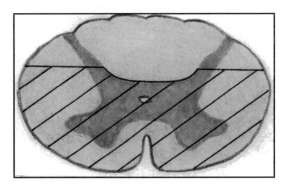

Etiology

- Infarct.
- Ischemia.
- Trauma.

Pathology

- Anterior Cord Syndrome Occurs when Two-Thirds of the Anterior Cord is Lost.

Symptomatology

- The Dorsal Columns are Spared.
    - Discriminative Touch, Vibration, Pressure, and Proprioception are spared
- The Lateral Corticospinal Tracts are Lost.
    - Because the Lateral Corticospinal Tracts descend down the Lateral White Funiculus, they are Lost
    - This results in Bilateral Spastic Paralysis
    - The patient loses Bilateral Voluntary Motor Control below the Level of the Lesion
- The Ventral Horn is Lost.
    - The Motor Neurons in the Ventral Horn are part of the LMN System
    - Because the Ventral Horn Motor Neurons are lost, the patient presents with Flaccidity at and below the Lesion Level
- The Spinothalamic Tracts are Lost.
    - Because the Spinothalamic Tracts synapse in the Dorsal Horn and Decussate across the Anterior White Funiculus as soon as they enter the cord, the patient presents with Bilateral Loss of Pain and Temperature

CENTRAL CORD SYNDROME
(also called Syringomyelia)

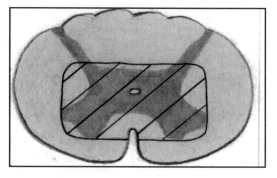

Etiology
- Unknown.

Pathology
- Central Cord Syndrome involves a Cavitation of the Central Cord in the Cervical Segments.

Symptomatology
- The Spinothalamic Tracts are the First to be Lost.
  - Because the Spinothalamic Tracts synapse in the Dorsal Horn and Decussate to the Anterior White Funiculus as soon as they enter the Cord, the Spinothalamic Tracts are lost
  - This results in Bilateral Loss of Pain and Temperature
- The Ventral Horn is Lost.
  - This results in Flaccidity of the Upper Extremities, because the disease occurs in the Cervical Regions of the Spinal Cord

POSTEROLATERAL CORD SYNDROME

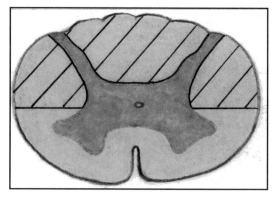

Etiology
- Degeneration of the Spinal Cord from Severe Vitamin B12 Deficiency, Pernicious Anemia, or AIDS.

Pathology
- Affects the Posterior and Posterolateral White Funiculi of the Spinal Cord.

Symptomatology
- The Dorsal Columns are Lost Bilaterally.
  - Results in Bilateral Loss of Discriminative Touch, Pressure, Vibration, and Proprioception
- The Lateral Corticospinal Tracts are Lost Bilaterally.
  - Because the Lateral Corticospinal Tracts Descend in the Lateral White Funiculus, they are lost
  - Results in Bilateral Spastic Paralysis
- The Spinocerebellar Tracts are Lost.
  - This results in Bilateral Ataxia

ANTERIOR HORN CELL SYNDROME
(also called Ventral Horn Syndrome)

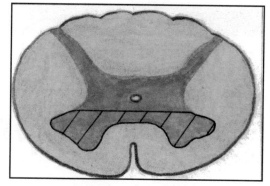

Etiology

- Disease process that Destroys the Motor Neurons in the Ventral Horn.

Pathology

- Anterior Horn Cell Syndrome involves LMN Damage.

Symptomatology

- Results in Bilateral Flaccidity in the Muscles Innervated by the Affected Spinal Cord Levels.

- An example is Poliomyelitis: An Acute Viral Disease affecting the Ventral Horn Motor Cell Bodies.

Proprioception

PROPRIOCEPTION

- Sherrington named the term Proprioception and defined it as the Sixth Sense: "The sense by which the body knows itself, judges with perfect, automatic, instantaneous precision the position and motion of all of its movable parts, their relations to one another, and their alignment in space."
- Proprioception is the ability to sense one's body position in space.
- The Latin root Proprius means to "own oneself" or to feel one's body as one's own.
- Proprioception occurs mostly on an unconscious level because it is primarily mediated by the Cerebellum.
- The term Kinesthesia refers to the ability to sense one's body Movement in space.

THE ABILITY TO SENSE ONE'S BODY IN SPACE IS BASED ON THREE SYSTEMS

Visual System

- Humans use Visual Cues to negotiate the environment.

Vestibular System

- The Labyrinthines in the inner ear—the semi-circular canals—contain continuously moving liquid that is constantly monitored by the Vestibular System to provide feedback regarding the Head's Position in Space.
- The Vestibular System also has neural connections to the Cerebellum to provide feedback about the Head's Position in Space.

Proprioceptive System

- The Proprioceptive System is a Feedback/Feedforward Loop between Muscle Spindles, the Golgi Tendon Organs, the Joint Receptors, and the Cerebellum. Together, these provide constant information about the body's position in space.

MUSCLE SPINDLE

- Muscle Spindles, or MSs, are Proprioceptors located in Skeletal Muscle.
- MSs are Sensory Receptors that provide a constant flow of information regarding Length, Tension, and Load on the muscles.
- MSs detect when a muscle has been stretched and initiate a reflex that resists the stretch.

Extrafusal Muscle Fibers

- Extrafusal Muscle Fibers are the Bulk of the Muscle.

Intrafusal Muscle Fibers

- Intrafusal Muscle Fibers are the MSs that sit within the Bulk of the Muscle.
- The Intrafusal Muscle Fibers are attached to the Extrafusal Muscle Fibers.

Density of Muscle Spindles in an Extrafusal Muscle Fiber

- The more MSs in an Extrafusal Muscle Fiber, the Greater Precision Control the Muscle has.
- Muscles with the Highest Density of MSs are small muscles designed for Fine Motor Control.

NUCLEAR CHAIN FIBERS AND NUCLEAR BAG FIBERS

- The Nuclear Chain and Nuclear Bag are structures within the Muscle Spindle.

Nuclear Bag

- The Nuclear Bag is responsive to Changes in Muscle Length.
- When the Length of the Muscle Changes, the Nuclear Bag Fires.
- The Nuclear Bag also detects the Velocity of the Muscular Stretch, or how quickly the Muscle Stretched.

Nuclear Chain

- The Nuclear Chain only Fires in response to a New Muscle Length—it does NOT respond to Velocity, like the Nuclear Bag.

Properties of the Equitorial Regions of the Nuclear Bag and Chain

- The Equitorial Region of the Bag is Elastic and Phasic (quick responding).
- The Equitorial Region of the Chain is Elastic and Tonic (slow responding).

Properties of the Polar Regions of the Nuclear Bag and Chain

- The Polar Regions of the Bag are Contractile and Tonic (slow responding).
- The Polar Regions of the Chain are Contractile and Phasic (quick responding).

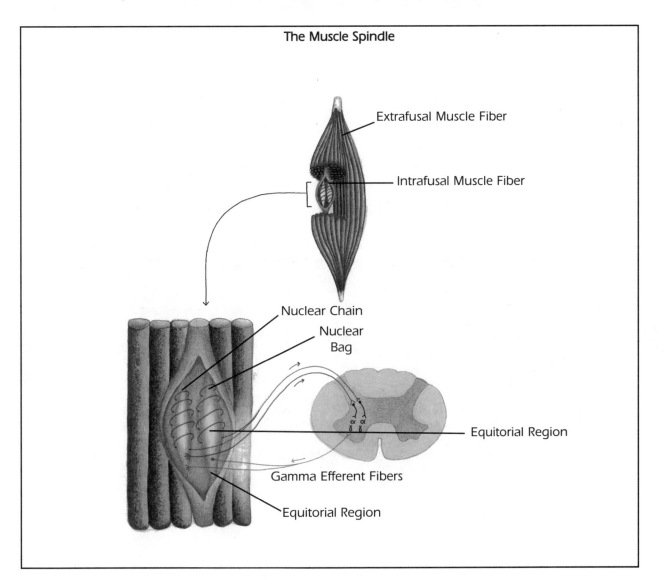

The Muscle Spindle

Extrafusal Muscle Fiber

Intrafusal Muscle Fiber

Nuclear Chain

Nuclear Bag

Equitorial Region

Gamma Efferent Fibers

Equitorial Region

THE MUSCLE SPINDLES USE TWO TYPES OF SENSORY FIBERS TO SEND INFORMATION TO THE VENTRAL HORN

Ia (Primary Ending)

- The Ia Sensory Fibers are Large and Heavily Myelinated.
- They are Fast Conducting.
- The Ia Fibers wrap around the Equitorial Region of both the Bag and Chain.
- The Ia Fibers respond to the Rate of Muscle Stretch (Velocity) and to Changes in Muscle Length.
- The Ia Fibers are Fast Adapting (Phasic).

II (Secondary Ending)

- II Fibers are Medium Size Fibers.
- They Terminate on the Equitorial Region of the Chain Only.
- II Fibers are predominantly located on the Chain.
- II Fibers respond to Changes in the Length of the MS (the rate of the stretch is not involved).
- II Fibers are Slow Adapting (Tonic).

MUSCLE SPINDLE SEQUENCE OF EVENTS

- The Extrafusal Muscle Fiber Stretches.
- This causes the MS (the Intrafusal Muscle Fibers) to Stretch.
- The Equitorial Region of the Bag stretches right away—because the Equitorial Region of the Bag is Elastic and responds more to Stretch than does the Polar Regions. (If the Stretch is a Sustained Stretch, the Equitorial Region of the Chain will also Stretch.)
- Stretching of the MS causes the Ia Fibers to Fire.
- The Ia Fibers will fire in response to a Quick or Phasic response—because the Ia fibers from the Bag are Phasic.
- The Ia Fibers will also fire in response to a Tonic or Sustained and Slow response—because the Ia Fibers from the Chain are Tonic.
- The Secondary Fibers, the II Fibers, then fire.
- The II is attached only to the Chain and the Chain only detects Length and Position Changes.
- The II Fibers are Tonic and respond to a Slow, Sustained Stretch.

PROPRIOCEPTIVE INFORMATION FROM THE MS GOES TO 4 PLACES

- Proprioceptive Information from the MS travels along the Ia and II Fibers to the Dorsal Horn.
- In the Dorsal Horn the Fibers synapse on an Interneuron and connect to Alpha Motor Neurons in the Ventral Horn.
- Information from the MS Travels to:
  - An Alpha Motor Neuron of the SAME Muscle (the Agonist) for facilitation of that Muscle. Referred to as Autogenic Excitation.
  - An Alpha Motor Neuron of the Antagonist Muscle for Inhibition of the Antagonist. Referred to as Reciprocal Inhibition: Every time an Alpha turns on an Agonist, an Alpha Turns off the Antagonist.
  - Renshaw Cells, or Special Interneurons that Modify the Action of Synergy Muscles. A Renshaw cell is a short axon that connects Motor Nerve Fibers with each other to produce Refined Motor Movement. Recurrent inhibition: Every time an Alpha turns on an Agonist, An Alpha also modifies (usually inhibits) the action of the Synergy Muscles.
  - The Cerebellum. MSs send Proprioceptive Signals to inform the Cerebellum of all Changes in Muscle Position and Length.

Information from the muscle spindle travels to four primary neuroanatomical structures.

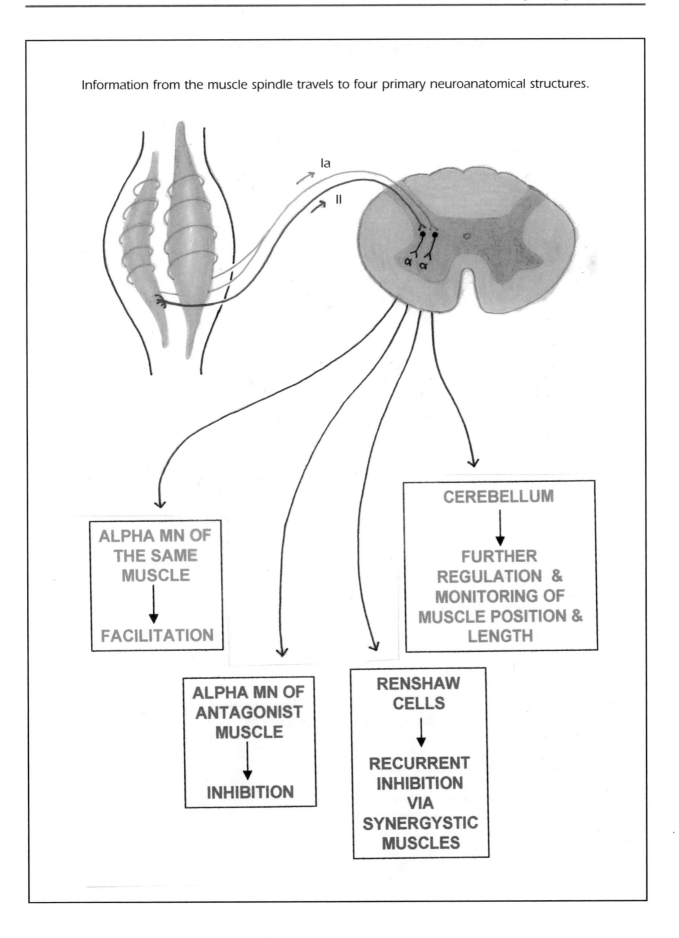

ALPHA MN OF
THE SAME
MUSCLE

↓

FACILITATION

ALPHA MN OF
ANTAGONIST
MUSCLE

↓

INHIBITION

RENSHAW
CELLS

↓

RECURRENT
INHIBITION
VIA
SYNERGYSTIC
MUSCLES

CEREBELLUM

↓

FURTHER
REGULATION &
MONITORING OF
MUSCLE POSITION &
LENGTH

THE GAMMA MOTOR NEURON SYSTEM

- The Gamma Motor Neurons are also located in the Ventral Horn, along with the Alpha Motor Neurons.
- The Gamma Motor Neurons send Proprioceptive Information from the Ventral Horn, back to the Muscle Spindle.
- Gamma Motor Neurons are Fast Conducting because they are Heavily Myelinated and Large.
- Gamma Motor Neurons innervate the Muscle Spindles.
- Alpha Motor Neurons innervate the Extrafusal Muscle Fibers (the muscle bulk).

Two Types of Gamma Motor Neurons

Gamma 1 Fibers
- Gamma 1 Fibers have plate endings and terminate on the Polar Regions of the Nuclear Bag.

Gamma 2 Fibers
- Gamma 2 Fibers have trail (or multibranching) endings and terminate predominately on the Nuclear Chain, adjacent to the Equitorial Region.

Gamma Motor Neuron Stimulation
- When the Gamma Motor Neurons in the Ventral Horn are stimulated, they fire causing the Polar Regions of the Bag and Chain to Contract.
- This causes the Equitorial Regions to Stretch (or Distort).
- The Ia Fibers then fire, causing the Alpha Motor Neurons in the Ventral Horn to fire.
- An Alpha Motor Neuron in the Ventral Horn innervates the Agonist, causing the Extrafusal Muscle Fiber to Contract.

GOLGI TENDON ORGANS

Location
- The GTOs are embedded in the Tendons, close to the Skeletal Muscle Insertions.

Function
- The GTOs are Proprioceptors that detect Tension in the Tendon of a Contracting Muscle.

Golgi Tendon Organs Use Ib Afferent Neurons
- The GTOs use Ib Afferent (or sensory) Neurons to send Proprioceptive Information to the Dorsal Horn.
- The Ib Fibers synapse on Interneurons in the Ventral Horn.
- In the Ventral Horn, the Interneurons synapse on Alpha Motor Neurons.

Autogenic Inhibition
- Activation of the GTOs causes a Contracting Muscle to be Inhibited; in other words, it relaxes.
- This is a Protective Function: If the GTOs did not become activated in response to a muscle's stretch, an individual could easily tear his or her muscles.

Sequence of GTO Events

1. The Agonist Muscle Contracts.
2. This Activates the GTOs (which are embedded in the Antagonist of the Contracting Muscle).
3. The GTOs send Proprioceptive Information along the Ib Sensory Fibers to the Dorsal Horn.
4. In the Dorsal Horn, the Ib Fibers synapse with an Interneuron.
5. The Interneuron connects with an Alpha MN in the Ventral Horn.
6. Information from the GTOs goes to 3 places:
   - An Alpha MN to Inhibit the Contracting Agonist Muscle: Autogenic Inhibition.
   - An Alpha MN to Facilitate the Antagonist of the Contracting Muscle.
   - The Cerebellum for further Proprioceptive Feedback.

GOLGI TENDON ORGAN FUNCTIONAL IMPLICATIONS

- The GTOs help to control the Speed of a Contraction for Coordinated, Fine, Precision Movements.
- Protective Mechanism: Humans need the action of the GTOs to Protect against Muscle Tears and Pulls.
- The GTOs help to Reduce Muscle Cramps:
  - When an individual has a cramp in a muscle group (eg, when the Plantar Flexors are cramping), placing a Sustained Stretch on that muscle group will kick in the GTOs to reduce the cramp. In other words, position the Plantar Flexors in a Sustained Stretch (ie, pull the Plantar Flexors into Dorsiflexion for a sustained stretch).
  - The GTOs in the Agonist become activated and inhibit the Cramping Agonist (the Plantar Flexors).
- Clasp Knife Syndrome. (Clasp Knife Syndrome occurs when a muscle group is hypertonic.)
- To reduce spasticity, perform a sustained stretch on the contracting spastic muscle.
- This will activate the GTOs and relax the hypertonic muscle.

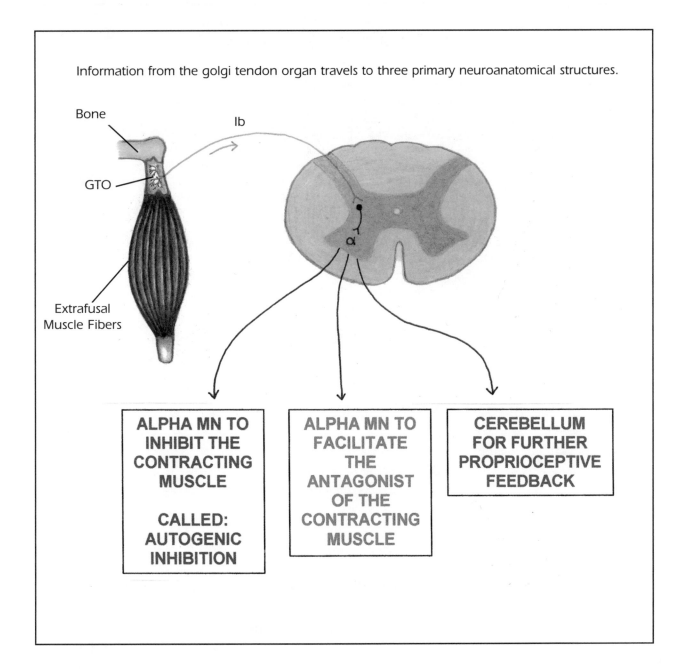

Information from the golgi tendon organ travels to three primary neuroanatomical structures.

Bone

Ib

GTO

α

Extrafusal
Muscle Fibers

ALPHA MN TO INHIBIT THE CONTRACTING MUSCLE

CALLED: AUTOGENIC INHIBITION

ALPHA MN TO FACILITATE THE ANTAGONIST OF THE CONTRACTING MUSCLE

CEREBELLUM FOR FURTHER PROPRIOCEPTIVE FEEDBACK

Neurologic Concepts Underlying the Basis of Many Therapeutic Techniques

- SLOW SUSTAINED STRETCHES will reduce spasticity in a muscle group by activating the GTOs; the GTOs will inhibit the Contracting Spastic Agonist and facilitate the Antagonist.
- SPLINTING a Spastic Muscle Group essentially positions the spastic limb in a Sustained Stretch. The GTOs become activated and inhibit the Spastic Contracting Muscle Groups (Agonists) while also facilitating the Antagonists.
- SERIAL CASTING also places a Spastic Contracting Muscle Group (Agonists) on a Sustained Stretch. This Activates the GTOs, causing inhibition of the Spastic Muscle Group and facilitation of the Antagonists.
- SEATING and POSITIONING TECHNIQUES are also used to place a Spastic Contracting Muscle group on a Sustained Stretch. Again, the GTOs will become activated, causing a reduction in the Spastic Muscle Groups and facilitating the Antagonist Muscle Groups.

Muscle Spindles and Gamma Motor Neurons Form a Feedback/Feedforward Loop

- There is some change in an Extrafusal Muscle Fiber (muscle stretch).
- The Intrafusal Muscle Fibers (the Nuclear Bag and Chain of the MS) then stretch.
- The Bag and Chain send Proprioceptive Information to the Alpha MNs in the Ventral Horn via the Ia and II fibers.
- This Proprioceptive Information from the MS goes to 4 places:
  - An Alpha MN of the Same Muscle for Facilitation of that Muscle (the Agonist): Autogenic Excitation.
  - An Alpha MN of the Antagonist Muscle for Inhibition of the Antagonist: Reciprocal Inhibition.
  - Renshaw Cells for Modification (usually inhibition) of Synergy Muscles: Recurrent Inhibition.
  - Cerebellum for further monitoring and modification of that muscle group's movement. The Cerebellum uses this Proprioceptive Information for the Fine Tuning of Precision, Coordinated Movement.
- Alpha Excitation also stimulates the Gamma MNs in the Ventral Horn.
- Gamma MNs fire and send Proprioceptive Information along the Gamma 1 and 2 Fibers to the MSs.
- This causes the Intrafusal Muscle Fibers (MSs) to contract.
- The MSs send their Proprioceptive Information to the Ventral Horn, via the Ia and II fibers.
- The Feedback/Feedforward system begins over again, sending information to the 4 places described above.

GAMMA BIASING

- The function of the Gamma Motor Neurons is to make the MS more sensitive and keep the muscles primed, or ready, for action.
- Gamma Biasing raises the level of firing of the MS.
- When activation of the Gamma Motor Neurons raises the level of MS firing, it causes the muscle to prepare, or set, for an anticipated activity.

Examples of Gamma Biasing
- Walking up a set of steps in which the step height is not uniform. An individual will raise his or her foot to the uniform step level and then actively move his or her foot downward too heavily on the step that is lower than anticipated.
- An individual lifts a box of supplies that he or she believes is much heavier than it actually is. The individual lifts the box too quickly and too high as a result of anticipating the box to be heavier. His or her muscles were primed, by Gamma Biasing, to lift a heavier box.

JOINT RECEPTORS

Location
- Joint Receptors are located in the Connective Tissue of a Joint Capsule.

Three Events Cause the Gammas to Fire

- The Supraspinal Motor Centers (Basal Ganglia, Vestibular Nuclei, Reticular Formation).
  - These Supraspinal Centers send projections to the Gammas in the Ventral Horn through Descending Pathways: Vestibulospinal and Reticulospinal Tracts
- Cutaneous Stimulation.
  - Sensory Information from the Skin is sent to the Dorsal Horn
  - This Information synapses on Interneurons and excites the Gamma MNs
- Alpha and Gamma Co-Activation.
  - Every time an Alpha MN is activated, a Gamma MN is co-activated

Function
- Joint Receptors respond to Mechanical Deformation occurring in the Joint Capsule and Ligaments.
- Joint Receptors send this Proprioceptive Information to the Cerebellum and to the Ventral Horn.

What Stimulates the Joint Receptors?
Ruffini's Endings
- Located in the Joint Capsule.
- Signal the Extremes of Joint Range.
- Respond more to Passive than Active Movement.

Paciniform Corpuscles
- Located in the Joint Capsule.
- Respond Only to Dynamic Movement; that is when the Joint is Moving.

Ligament Receptors
- Similar in Function to GTOs.
- Located in the Ligaments of a Joint Capsule.
- Respond to Tension in the Joint Capsule.

Free Nerve Endings
- Located in the Joint Capsule.
- Most often Stimulated by Inflammation.
- Signal the Detection of Pain in a Joint.

Joint Receptors Use Several Different Sensory Fibers to Send Proprioceptive Information to the Cerebellum and Ventral Horn
- Ligament Receptors use Ib Sensory Fibers.
- Ruffini's Endings and Paciniform Corpuscles use II Sensory Fibers.
- Free Nerve Endings use Delta A and C Sensory Fibers.

Joint Receptors Send Proprioceptive Information to Three Places

- The Joint Receptors use sensory fibers to send Proprioceptive Information to the Dorsal Horn. Here the messages synapse on an Interneuron.
- The Interneuron connects with a Motor Neuron in the Ventral Horn and sends Proprioceptive Messages back to the Muscles surrounding the Joint.
- An Alpha MN in the Ventral Horn also sends Joint Receptor Information to the Cerebellum for constant feedback about Joint Position and Movement.

While Muscle Spindles, Golgi Tendon Organs, and Joint Receptors are critical for normal Proprioception, Joint Receptors alone may not be essential for Proprioception.

This is suggested because individuals with Joint Replacements still retain good Proprioception in Joint Midrange.

Disorders of Muscle Tone

MUSCLE TONE

- Muscle Tone is a Continuous State of Muscle Contraction at Rest.
- Tone is an Unconscious Phenomenon—humans cannot consciously will muscles to increase or decrease in tone.

PRIMARY MECHANISMS THAT MEDIATE TONE

Extrapyramidal Structures (Part of the Upper Motor Neuron System)

- The Extrapyramidal Structures are Motor Centers and Pathways Located outside of the Pyramidal System.
- These include Brainstem Centers such as the Vestibular Nuclei and the Reticular Nuclei.
- Also include Extrapyramidal Motor Tracts such as the Vestibulospinal, Rubrospinal, and Reticulospinal Pathways.

Basal Ganglia

- Including the Caudate, Putamen, and Globus Pallidus (of the UMN System).

Pyramidal Structures

- Includes the Corticospinal Tracts (of the UMN System).

Cerebellum

- The Cerebellum works in a Feedback/Feedforward Loop with the Brainstem, Basal Ganglia, Extrapyramidal Structures, and the Muscle Spindles, Golgi Tendon Organs, and Joint Receptors.
- The Cerebellum also mediates Tone through the Afferent Sensory Tracts that travel to the Cerebellum: Posterior and Anterior Spinocerebellar Tracts, Cuneocerebellar Tracts, and the Rostral Spinocerebellar Tracts.

Motor Neurons of the Ventral Horn (of the Lower Motor Neuron System)

- The Alpha and Gamma Motor Neurons of the Ventral Horn mediate Tone.

Peripheral Nerves that Innervate Skeletal Muscle (of the Lower Motor Neuron System)

CLASSIFICATIONS OF TONE

Hypotonicity

- Hypotonicity is an abnormal decrease in muscle tone; eg, Floppy Babies, Individuals with spinal cord injury (SCI) below L1.
- Hypotonicity is caused by LMN Lesions:
    - Damage to the Motor Neurons in the Ventral Horn.
    - Damage to the Spinal Nerves in the Periphery.
- Lesions to the Posterior Cerebellar Lobe (Neocerebellar Lobe) produce Hypotonicity and Hyporeflexia.

Hypertonicity

- Hypertonicity is an abnormal increase in muscle tone, accompanied by resistance to active and passive movement.

- Hypertonicity is caused by UMN damage.
- Lesions to the Anterior Cerebellar Lobe (Paleocerebellar Lobe) produce Hypertonicity and Hyperactive Reflexes.

Spasticity

- Spasticity is a form of Hypertonicity.
- Spasticity involves difficulty actively and passively moving the spastic muscle groups on one side of a joint.
- Either the Flexors or Extensors are Spastic, but not both.
- Spasticity is associated with such disorders as SCI (T12 and above), Head Injury, Cerebrovascular Accident, and Cerebral Palsy.

Rigidity

- Rigidity is a form of Hypertonicity.
- Rigidity involves difficulty Actively and Passively moving the muscle groups on BOTH sides of a joint.
- Rigidity is associated with such disorders as Parkinson's disease.

Clasp Knife Syndrome

- Clasp Knife Syndrome involves severe rigidity at a joint.
- A Sustained Stretch will relax the muscle group and the rigidity will suddenly give way.

Cogwheel Rigidity

- In Cogwheel Rigidity, the resistance is jerky and characterized by a pattern of release/resistance in a quick jerky movement.

Clonus

- Clonus is an Uncontrolled Oscillation of a Muscle that occurs in a Spastic Muscle Group (results from UMN Lesions).

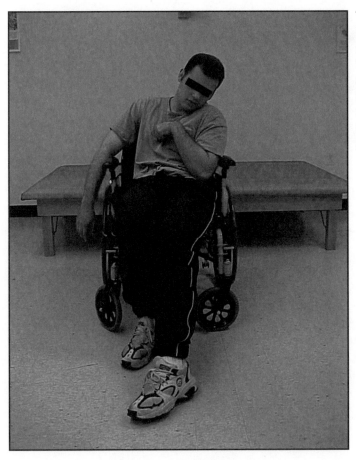

Typical posture of an individual with Left Hemiplegia. Increased tone on the contralateral side of the body. Head and Neck are flexed; Scapula is retracted and depressed; Shoulder is adducted and internally rotated; Elbow, Wrist, and Fingers are flexed; Forearm is pronated; Wrist is ulnarly deviated; Fingers are adducted; Hip, Leg, Knee, and Ankle are extended; Pelvis is posteriorly tilted; Thigh is internally rotated; Foot is inverted.

THEORIES OF SPASTICITY (ETIOLOGY)

Hyperactive Reflex Arcs
- The Reflex Arc is firing without modification from the Cortex.
- This occurs in an UMN SCI when the Corticospinal Tracts are Lost.
- Below the Lesion Level, the Corticospinal Tracts are lost, however, the Reflex Arc remains intact and fires without modification from the higher centers.
- This results in Spasticity.

Reduced Reciprocal Inhibition of the Antagonist and Synergy Muscles
- When Lesions cause a Reduction in the ability of the Alpha Motor Neurons to inhibit the Activity of Antagonist and Synergy Muscle Groups, the Agonist fires without modification.
- This can result in Spasticity.

Loss of the Cortical Modification of the Alpha Motor Neurons in the Ventral Horn
- When Lesions to the Brainstem and Subcortical Motor Center occur, Cortical Inhibition of these Motor Centers is lost.
- This causes the Alpha and Gamma Motor Neurons to fire without Modification from the Cortex.
- This can result in increased Tone.

Damage to the Primary Motor Area
- When damage to the Primary Motor Area occurs, the Corticospinal Tracts fire without Cortical Modification, causing Spasticity.

Damage to the Brainstem Regions that Contain the Supraspinal Motor Centers
- The Supraspinal Motor Centers include the Vestibular Nuclei, Reticular Nuclei, and Pontine Nuclei.
- When these are damaged, severe Spasticity can occur in the form of Extensor Tone.

THERAPEUTIC TECHNIQUES TO INFLUENCE TONE

Sustained Stretch on the Agonist
- Any Sustained Stretch on an Agonist (on a Spastic Contracting Muscle Group) will activate the GTOs, located in the Spastic Agonist.
- The GTOs serve to inhibit the Spasticity of the Agonist and facilitate the Antagonist Muscle Group.
- Example: If the Biceps are Spastic, place the elbow joint on a sustained stretch (in Extension). This will activate the GTOs; the GTOs will inhibit the Spastic Contracting Agonist and facilitate the Antagonist Muscle Groups.

Quick Stretch on the Agonist
- A Quick Stretch on the Agonist activates the MS.
- The MS sends messages to the Alpha Motor Neurons in the Ventral Horn to continue innervating the Agonist.
- Example: In a floppy baby, Quick Stretches on the Biceps may increase tone in the Biceps for Functional Upper Extremity use.

Placing Pressure on the Tendon of the Agonist
- Placing Pressure on the Tendon of the Agonist of a Spastic Muscle Group will activate the GTOs.
- The GTOs will inhibit the Spastic Contracting Agonist and facilitate the Antagonist.
- Example: If the Biceps are Spastic, place sustained pressure on the Tendons of the Biceps to activate the GTOs. This will relax the Spasticity in the Biceps and facilitate activity in the Triceps.
- NEVER place Pressure on the Muscle Belly of the Contracting Spastic Muscle. This will activate the MS and continue to Increase Spasticity in the Spastic Muscle.

Splinting

- Splinting works on the premise of positioning the Spastic Muscle Group on a Sustained Stretch to Reduce Tone in the Agonist.
- Example: Spasticity in the Wrist Flexors.
- Use a Resting Pan Splint to place the Wrist Flexors on a Sustained Stretch in Extension.
- The GTOs will become activated and will serve to reduce the Spasticity in the Agonist and facilitate activity in the Antagonist (the Wrist Extensors).
- It is very important to make sure that the Splint is well-fitting, or the Cutaneous Receptors will activate the Gamma Motor Neurons. The Gamma Motor Neurons will send signals to the MS to fire, thus facilitating Spasticity in the Agonist (the Wrist Flexors).

Serial Casting

- Serial Casting involves placing the Spastic Muscle Groups around a joint in a Cast to increase Range of Motion (ROM) over time.
- Example: The Elbow Flexors are Spastic.
- In the Initial Cast, the therapist pulls the Elbow Flexors into slightly greater Extension and places a Cast on the Elbow Joint in this position.
- Essentially, the Cast places the Elbow Flexors on a prolonged sustained stretch.
- The GTOs become activated and reduce the Spasticity in the Elbow Flexors.
- Every few days or once a week, the therapist places a new cast on the Spastic Joint.
- With each Successive cast, the Elbow Joint is positioned in greater Extension to increase ROM gradually.
- A common problem with Serial Casting is Skin Breakdown. If Skin Breakdown occurs, the cast must be removed and the skin must heal until any further casting can be done. Unfortunately, while the skin is healing, the patient's Spastic Muscle Group will often resume its initial degree of Spasticity. Any ROM that has been gained as a result of Serial Casting is often lost.

Reducing Clonus Through Sustained Stretching

- Clonus is an Uncontrolled Oscillation of a Muscle that occurs in a Spastic Muscle Group.
- Clonus can frequently be observed in the ankle joint of individuals with quadriplegia.
- When a therapist places the patient's foot on the wheelchair footplate, the therapist is essentially giving a Quick Stretch to the Plantar Flexors.
- This Quick Stretch activates the MS and causes the Plantar Flexors to Contract Uncontrollably.
- To Reduce Clonus:
  - Position the Plantar Flexors on a Sustained Stretch; in other words, pull the foot into Dorsiflexion on a Sustained Stretch.
  - This will activate the GTOs. The Spastic Muscle Group (the Plantar Flexors) will relax.

Ashworth Scale: Used to Evaluate Tone

1. NORMAL TONE
2. SLIGHT HYPERTONUS: Noticeable Catch when the Limb is being Passively Moved.
3. MORE MARKED HYPERTONUS: A Catch can be detected, but the Limb is Still able to be Passively Moved with ease.
4. MODERATE HYPERTONUS: Passive Movement of the Limb is Difficult to Achieve.
5. SEVERE HYPERTONUS: Passive Movement of the Limb is unable to be Achieved.

Motor Functions and Dysfunctions of the Central Nervous System:
Cortex, Basal Ganglia, Cerebellum

CEREBRAL CORTEX

Cortical Mapping of the Brain

- The first neuroanatomist to attempt brain mapping was Brodmann. He numbered each area of the Cortex by number.
- In 1909 he mapped 52 brain areas.
- Brodmann believed that each numbered area corresponded to a precise brain function.
- The boundaries between each area were often not precise.
- And the correlation between Brodmann's Areas and a specific function is not precise.

PET and MRI Scans

- Today PET Scans (Positron Emission Tomography) and MRIs (Magnetic Resonance Imaging) are used to more precisely map brain areas.

Functional Divisions of the Cerebrum

Archicortex

- The Archicortex is the Core of the brain. It is comprised of the Hippocampal Formation.
- The Hippocampal Formation is located deep in the brain. It is involved in Learning and Memory.
- The Archicortex is believed to be phylogenetically ancient.

Paleocortex

- The Paleocortex are the outer layers that sit over the core. It includes the Parahippocampal Gyrus.
- The Parahippocampal Gyrus is a region of the Limbic System that is adjacent to the Hippocampus.
- The Parahippocampal Gyrus relays information between the Hippocampus and other brain regions.

Neocortex

- The Neocortex is considered to be the newest part of the brain phylogenetically.
- It is comprised of the most superficial layers of the brain.
- Includes the Primary Motor and Sensory Cortices and the Association Cortices.
- The Neocortex is found only in Mammals. It makes up 50% to 80% of the brain's total volume.

THE CEREBRUM'S ROLE IN MOTOR CONTROL

Primary Motor Area (M1)

- The Primary Motor Area is the Precentral Gyrus. This is where Voluntary/Conscious Movement is initiated.

- The Corticospinal Tracts originate here.
- It is also the site of the Motor Homunculus.

Motor Homunculus

- The Motor Homunculus is the map that denotes each body part's Cortical Representation for Voluntary Movement.
- The Face and Mouth have a large amount of cortical representation. This is for the purpose of Speech and Eating.
- The Hands have a large cortical representation. This is for Fine Motor Control necessary to explore the environment.
- The Cortical Representation for the Lower Extremities is located in the Medial Longitudinal Fissure.

Organization of the Motor Homunculus

- The Organization of the Motor Homunculus is NOT PERMANENT.
- One needs to use the body part or the cortical representation for that body part disappears—as with amputees over time.

Motto of the Brain: "IF YOU DON'T USE IT, YOU'LL LOSE IT."

- Examples are a Peripheral Nerve Injury in which the hand becomes denervated for a length of time or an Orthopedic Injury causing prolonged inactivity of a limb.
- The Cortical Representation for these body parts will be rapidly appropriated by other body regions.
- Cortical maps are Use Dependent.
- Once a cortical area representing a body part is gone, the individual must recreate a new area through new experiences.

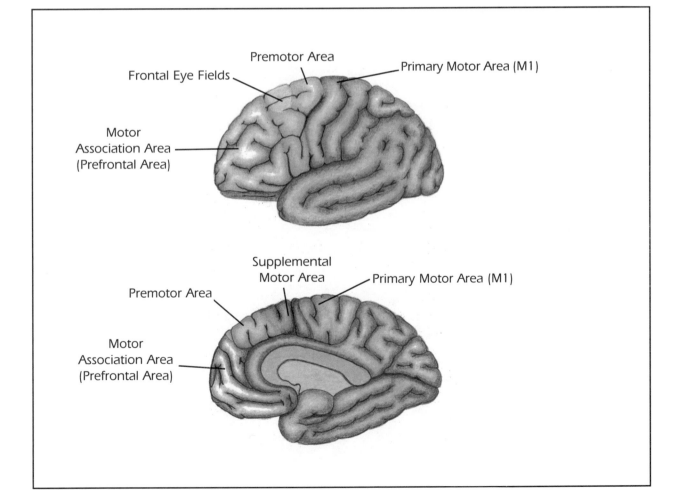

Lesion to the Primary Motor Area

- Loss of Voluntary Movement to the Contralateral Body Part.
- The Loss of Voluntary Movement in a Body Part corresponds to the Area of the Motor Homunculus that was damaged.
- A Lesion to the Primary Motor Area may also result in a loss of the ability to Implement a Specific Motor Plan.

Premotor Area

- The Premotor Area is located just anterior to the Primary Motor Area.
- The Premotor Area has a role in Motor Planning, or Praxis.

Lesion to the Premotor Area

- Apraxia, or Motor Planning Difficulties.
- Either the inability to understand the demands of the task.
- Or the inability to access the appropriate Motor Plan.

Motor Association Area

- Also called the Prefrontal Area.
- Located in the Anterior Frontal Lobe.
- The Motor Association Area has a role in the Cognitive Planning of Movement.

Lesion to the Motor Association Area

- Loss of the Storage of Motor Plans.
- Inability to Cognitively think about how to carry out a motor task that was once known before injury/disease.
- The Premotor Area may be able to take over and Compensate for Damage to the Motor Association Area.

Supplementary Motor Area

- The Supplementary Motor Area is part of the Premotor Area.
- Located inside the Medial Longitudinal Fissure.
- Has a role in the Bilateral Control of Posture.

Lesion to the Supplementary Motor Area

- May result in Loss of Bilateral Control of Posture.
- Other areas may take over and Compensate for this loss—Cerebellum, Vestibular System.

Frontal Eye Field

- Located in the Middle Frontal Gyrus, anterior to the Premotor Area.
- Responsible for Visual Saccades.

Lesion to the Frontal Eye Field

- Results in Deviation of the Eyes to the same side of the Lesion.

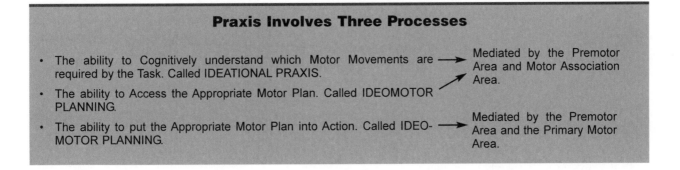

Praxis Involves Three Processes

- The ability to Cognitively understand which Motor Movements are required by the Task. Called IDEATIONAL PRAXIS. → Mediated by the Premotor Area and Motor Association Area.

- The ability to Access the Appropriate Motor Plan. Called IDEOMOTOR PLANNING.

- The ability to put the Appropriate Motor Plan into Action. Called IDEO-MOTOR PLANNING. → Mediated by the Premotor Area and the Primary Motor Area.

BASAL GANGLIA'S ROLE IN MOTOR CONTROL

Basal Ganglia

- The Basal Ganglia have a role in Stereotypic and Automated Movement Patterns (eg, walking, riding a bike).
- The Basal Ganglia structures include the Caudate, the Putamen, and the Globus Pallidus.

Divisions of the Basal Ganglia

Neostriatum

- Considered to be the newest region of the Basal Ganglia phylogenetically.
- Includes the Caudate and the Putamen.
- The Putamen is an Excitatory Structure.
- The Caudate has a role in the Inhibitory Control of movement. The Caudate acts like a Brake on certain motor activities.
- When that Brake is gone, Hyperkinetic Movement Disorders can occur—like Tics, Tourettes.

Paleostriatum

- Considered to be the older part of the Basal Ganglia phylogenetically.
- Includes the Globus Pallidus—an Excitatory Structure.

Corpus Striatum

- The Corpus Striatum is the Collective Name for all three structures: the Caudate, Putamen, and Globus Pallidus.

Collective Functions of the Basal Ganglia

Stereotypic Movement

- Stereotypic Movements are movement patterns that are "hard-wired" (ie, do not have to be learned on a conscious level) and develop normally as the individual's NS matures (eg, stretching the arms during yawning, using a reciprocal arm swing when walking).

Automated Movements

- Automated Movements are movement patterns that used to be mediated by conscious cortical control, but have been assumed by the Basal Ganglia structures once learned.
- This process underlies the saying, "Once you learn to ride a bike, you never forget," unless damage is sustained to the Basal Ganglia.
- Automated Movements include driving, riding a bike, and writing.

The Basal Ganglia Also Influence Tone

- The Basal Ganglia send input to the Alphas and Gammas in the Ventral Horn to influence Muscle Tone.

Basal Ganglia Pathways

Afferent Basal Ganglia Pathways (Leading to the BG)

- The Primary Motor Area and the Premotor Area send motor signals to the Caudate, Putamen, and Globus Pallidus.

Lenticular Formation (Traveling from the BG)

- This pathway travels from the Globus Pallidus to the Subthalamus, to the Reticular Formation, and to the Substantia Nigra.

Ansa Lenticularis (Traveling from the BG)

- This pathway travels from the Globus Pallidus to the Ventrolateral Nucleus of the Thalamus, to the Primary Motor area, and to the Premotor Area.

Nigrostriatal Pathway (Feedback Circuit of the BG)

- This pathway travels from the Substantia Nigra to the Globus Pallidus and Subthalamus.
- This circuit modifies the Ansa Lenticularis Pathway.
- In Parkinson's, the Substantia Nigra degenerates and the modification on the Ansa Lenticularis is lost.

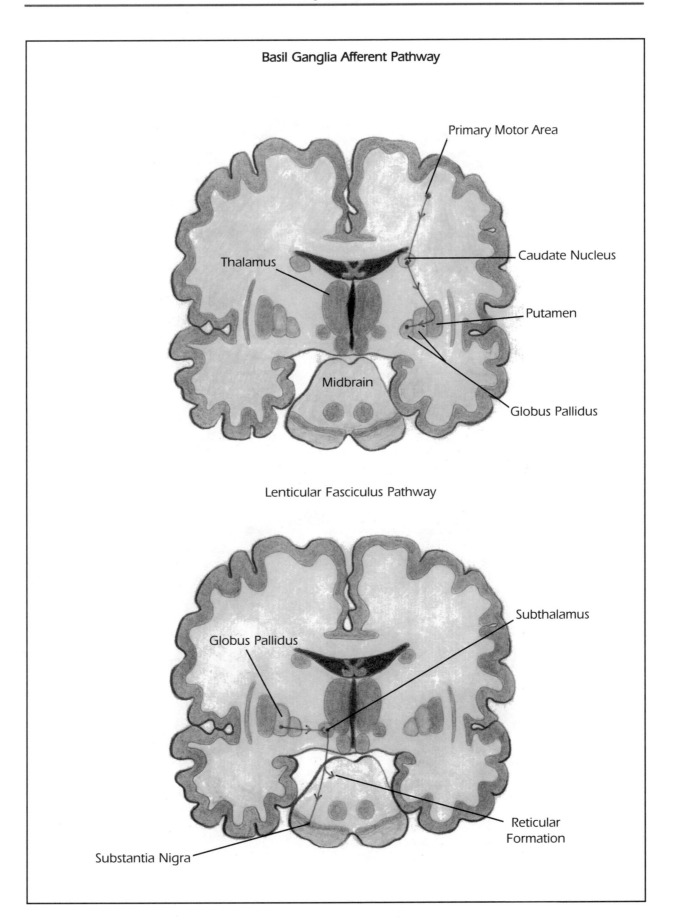

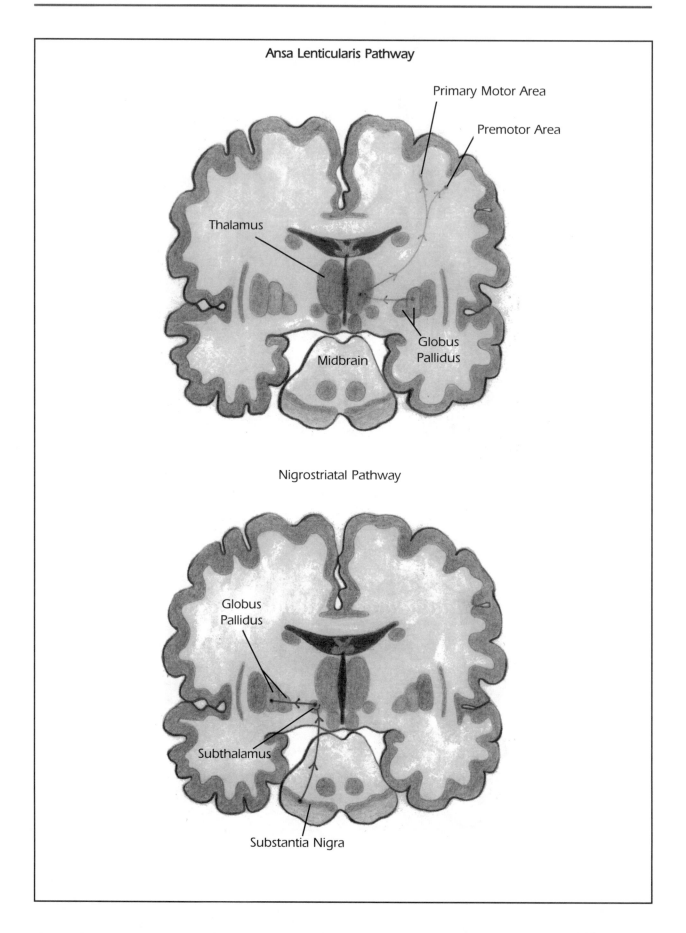

Ansa Lenticularis Pathway

Primary Motor Area

Premotor Area

Thalamus

Globus
Pallidus

Midbrain

Nigrostriatal Pathway

Globus
Pallidus

Subthalamus

Substantia Nigra

BASAL GANGLIA LESIONS

Basal Ganglia Lesions Present As

- Difficulty Initiating, Continuing, or Stopping Movement.
- Problems with Muscle Tone (particularly Rigidity).
- Increased Involuntary, Undesired Movements (Tremor, Chorea).

Hemiballismus

- Hemiballismus is a lesion of the Subthalamus and Caudate causing a disinhibition of neuronal activity between the Thalamus and the Cortex.
- Results in Violent Thrashing of the Contralateral Extremity.
- Ballismus is the name of the disorder if the lesion is bilateral.

Athetosis

- Involves a Slow Flailing of the Upper and Lower Extremities and Continuous Movement.
- Slow, Aimless, Purposeless Movement—especially of the Distal Musculature.
- Caused by a lesion to the Putamen and Caudate.
- Some Cortical Involvement.
- Athetosis is a Slower Movement than Chorea.

Chorea

- Choreiform means to Dance. Chorea is Sudden, Involuntary, Jerky Movements that look Dance-Like in quality.
- Usually involves Axial and Proximal Limb Areas: Shoulder Shrugs, Moving Hips, Crossing and Uncrossing the Legs.
- May involve the Face and Tongue (Grimacing and Tongue Protrusions).
- Caused by a lesion to the Caudate and Putamen.

Huntingdon's Chorea

- Huntingdon's Chorea is a Chronic Degenerative Disorder that also involves a Progressive Dementia.
- It is an Inherited Disease.
- Involves Degeneration of the Caudate Nucleus and some Areas of the Cortex.
- GABA Levels are diminished in the Striatum and Globus Pallidus.
- GABA is a Neurotransmitter that runs in the Nigrostriatal Pathway (from the Striatum to the Globus Pallidus).
- GABA in this pathway modulates the outflow of Dopamine—which runs in the Nigrostriatal Pathway.
- Increased Dopamine in the Basal Ganglia is believed to cause Huntingdon's Disease.

Dystonia

- Dystonia involves muscle contractions producing twisting movements that are repetitive and often result in abnormal postures.
- The sustained muscle contractions of a Dystonia can last from seconds to hours.

Parkinson's Syndrome

- Damage to the Nigrostriatal Pathway between the Substantia Nigra and the basal Ganglia.
- Dopamine is the Neurotransmitter that uses this pathway.
- In Parkinson's, the Substantia Nigra degenerates.
- This leads to decreased Dopamine available for the Basal Ganglia.
- Causes decreased modification of the Ansa Lenticularis Pathway (the pathway that modifies the Basal Ganglia Activity).

Parkinson's Symptoms

- Hypertonicity
- Cogwheel Rigidity
- Bradykinesia (poverty of movement)
- Mask-like Face (facial muscles are affected by Bradykinesia)
- Tremors at Rest (pill-rolling)
- Cunctation-Festinating Gait: Cunctation means to resist movement. Festination means to hurry. The individual experiences difficulty initiating and stopping movement.

Glossary of Terms of Idiopathic (no known origin) Torsion Dystonia

Involving the Eye

Blepharospasm: Eyes are involuntarily kept closed.

Oculogyric Crises: Attacks of forced deviation of gaze, often associated with a surge of Parkinsonism, catatonia, tics, and obsessiveness.

Opathalomoplegia: Paralysis of gaze.

Involving the Throat, Jaw, Lips, or Tongue

Oromandibular Dystonia: Involuntary opening and closing of the jaw, retraction or puckering the lips.

Lingual Dystonia: Repetitive protrusion of the tongue or upward deflection of the tongue toward the hard palate.

Laryngeal Dystonia: Speech is tight, constricted, and forced. Smooth flow of speech is lost.

Pharyngeal Dystonia: Associated with dysphagia, dysphonia (hoarseness), and dysarthria (slurred words).

Involving the Neck

Torticollis: Dystonic contractions of the neck muscles.

Retrocollis: Neck forced backward into hyperextension.

Anterocollis: Neck forced forward into hyperflexion.

Laterocollis: Neck forced laterally.

Involving the Trunk

Truncal Dystonia: Manifests as a lordosis, scoliosis, kyphosis, tortipelvis, opisthotonos (forced flexion of the head on the chest).

Involving the Legs

Crural Dystonia: Dystonic movement of the legs.

Dystonia of Single or Multiple Body Parts

Focal Dystonia: Dystonia in a single body part.

Multifocal Dystonia: Dystonia of more than one body part.

Hemidystonia: Involvement of limbs on one side of the body (due to a space occupying lesion or infarction).

Generalized Dystonia: Dystonia in a leg, the trunk, and one other body part; or both legs and the trunk.

As with most hyperkinetic movement disorders, the abnormal movements usually disappear during sleep and worsen with anxiety, fatigue, temperature changes, and pain.

- Loss of Arm Swing while walking
- Impaired Postural Reflexes (difficulty righting oneself and maintaining balance)
- Micrographia (microscopic handwriting)

Pharmacologic Treatment for Parkinson's

- L Dopa has been the most common treatment for Parkinson's (L Dopa Derivatives).
- L Dopa is a precursor to Dopamine—Dopamine cannot cross the Blood-Brain Barrier.
- When L Dopa is used and it crosses the Blood-Brain Barrier, it is synthesized into Dopamine.
- With the increase in Dopamine, the Nigrostriatal Pathway can run properly and modify the Ansa Lenticularis.
- A common problem is that the patient becomes less responsive to the drugs over time.

Kinesia Paradoxa in Parkinson's

- Kinesia Paradoxa is the sudden total conversion of Parkinsonism to normality or to hyperkinesia due to pharmacologic treatment.

Surgical Procedures for Parkinson's

- A Pallidotomy involves the removal of the Globus Pallidus.
- This a last resort treatment option.

Tremor

- Involuntary oscillating movement resulting from alternating or synchronistic contractions of opposing muscles.
- There are different types of Tremor.

Essential Tremor

- Most common form of Tremor.
- Characterized by Tremor upon movement (not a resting Tremor as in Parkinson's).
- Fingers and Hands are affected first.
- Tremor moves proximally to the Head and Neck.
- Slow Progressive Disorder.

Tics

- Tics are repetitive, brief, rapid, involuntary, and purposeless movements involving single or groups of muscles.
- Tics are fragments of movements or thoughts that are split off from more integrated behavior.
- People with Tics describe an Inner Tension that builds. This inner tension can be relieved by the ticing behavior.
- Tics have a conscious component. The conscious experience of feeling compelled to make sound may imply that Tics are a form of compulsion.
- A Tic can involve a brief isolated movement like eye blinks, head jerks, or shoulder shrugs.
- A Tic can be a variety of sounds: throat clearing, grunting.
- A Tic can also be associated with Intrusive Thoughts that the individual feels compelled to express.
- Tics can be meaningful utterances.
- Echolalia is an involuntary repetition of words just spoken by another person.
- Palilalia is an involuntary repetition of words or sentences.
- Coprolalia is an involuntary utterance of curse words.

Etiology of Tics

- The Dopamine System is a form of modifying and modulating movement.
- Increased sensitivity to Dopamine in the Basal Ganglia causes severe Tics.
- People with a normal response to Dopamine have mild Tics—this explains why normal people sometimes have Tics or mild obsessional thoughts, especially with stress.
- With increased sensitivity to Dopamine, the Cuadate (which normally acts like a Brake on certain Extraneous Movements) cannot suppress movements like Tics.

Tourette's Syndrome

- Tourette's Syndrome is a type of movement disorder involving tics.
- 70% of people with Tourette's Syndrome also have Obsessive Compulsive Disorder (OCD).
- OCD may represent an alternative expression of Tourette's Syndrome.
- The cause of Tourette's and OCD is likely a Dopamine and Basal Ganglia Disorder.
- 60% of people with Tourette's have Attention Deficit Hyperactivity Disorder (ADHD) involving short attention span, restlessness, poor concentration, diminished impulse control, and hyperactivity.
- A high percentage of people with Tourette's have concomitant learning disabilities, aggressiveness, anxiety, panic disorder, depression, mania, conduct disorder, phobias, dyslexia, and stuttering. These disorders have been found 5 to 20 times more commonly in people having Tourette's than in the general population.
- Tourette's also involves a defect in Motor Pattern Generators.
- Motor Pattern Generators are found in the Brainstem and Cortex.
- These produce a variety of hard-wired movements—reaching, grasping, walking, standing upright.
- These movement patterns are involved in all body postures and are present at birth; they are ready to mature as an infant begins to move.
- Increased sensitivity to Dopamine causes faulty inhibition of Motor Pattern Generators.

Tardive Dyskinesia

- Tardive Dyskinesia is a movement disorder related to treatment with Dopamine Receptor Antagonists (Neuroleptics and Antiemetics).
- The Term Tardive refers to the fact that this movement disorder occurs after chronic use of these drugs.
- Characterized by choreiform movements, dystonia, tics, and/or myoclonus.
- Example: Tongue Protrusions (orobuccolingual movements), Chewing-Type Movements, Facial Grimacing, Blepharospasm, Lip Smacking.
- Tardive Dyskinesia is different from most disorders because the discontinuation of the causative agent (the neuroleptic) does not result in the amelioration of the movement disorder.

Pathology of Tardive Dyskinesia

- The chronic use of neuroleptics results in a chronic blockage of Dopamine Receptors in the Basal Ganglia.
- The sensitivity of the Dopamine Receptors becomes severely decreased.
- Also suspected is a decrease in the Neurotransmitter GABA in the Basal Ganglia.

THE CEREBELLUM'S ROLE IN MOTOR CONTROL

Cerebellum

- The Cerebellum has a role in the Coordination of Movement, the Maintenance of Posture, and Equilibrium.
- Its major function is Proprioception.
- The Cerebellum could be called an "Error-Correcting Device" for the Motor System. It receives Proprioceptive information from the body and sends back information to modify muscle and joint activity for the achievement of Precision Motor Control.

Three Cerebellar Lobes

Archicerebellum (Flocculonodular Lobe)

- Receives input from the Vestibular Nuclei through the Inferior Cerebellar Peduncle.

Paleocerebellum (Anterior Lobe)

- Receives the Posterior and Anterior Spinocerebellar Tracts.
- Influences muscle tone by sending efferent fibers to the Vestibular Nuclei and the Reticular Formation, through the Superior Cerebellar Peduncle.

Neocerebellum (Posterior Lobe)

- Receives information from the Cerebral Hemispheres.
- Information from the Cortex descends to the Pontine Nuclei and then travels to the Cerebellum through the Middle Cerebellar Peduncle.

CEREBELLAR LESIONS

Neocerebellar Lesions

- Neocerebellar lesions present with the following:
- IPSILATERAL ATAXIA: The Posterior Lobe receives input from the Cortex. The Dorsal Columns mediate Conscious Proprioception. Loss of this Cortical Input results in Ataxia.
- IPSILATERAL HYPOTONIA and HYPOREFLEXIA
- DYSMETRIA: Inability to judge distance. Past-pointing or over-shooting one's grasp of objects occurs. People with Dysmetria use Visual Cues for the readjustment of imprecise movement.
- ADIADOCHOKINESIA: Inability to perform rapid alternating movements.
- INTENTION TREMORS: These occur During movement, as opposed to Resting Tremors—which occur in Parkinson's Disease.
- REBOUND PHENOMENON: Inability to regulate reciprocal movements. Example: The therapist gives resistance to the patient's flexed arm. When the therapist releases her resistance, the patient has no control or regulation over his speed of movement and his hand hits his chest. This occurs because immediate proprioceptive feedback is lost—the individual cannot regulate the speed of his arm movement quickly enough.

- ATAXIC GAIT: The Posterior Lobe receives input from the Cortex. The Dorsal Columns mediate conscious proprioception. Loss of this Cortical Input results in an Ataxic Gait.
- STACCATO VOICE: Broken Speech. The modulation of speech is a proprioceptive function.

Paleocerebellar Lesions

- Lesions in the Anterior Lobe produce severe DISTURBANCES IN EXTENSOR TONE.
- This is because this lobe receives the Spinocerebellar Tracts.
- When the Spinocerebellar Tracts are lost, an increase in Extensor Tone results.

Archicerebellar Lesions

- Lesions of the Flocculonodular Lobe result in Uncoordinated Trunk Movements—ATAXIA.
- The Flocculonodular Lobe receives input from the Vestibular Nuclei, the Cuneocerebellar Tract, and the Rostral Cerebellar Tract.
- Loss of this Vestibular Input results in Balance Deficits—particularly in the Trunk and Upper Extremities.
- NYSTAGMUS also occurs because the Cerebellum has connections to the Oculomotor System via the Medial Longitudinal Fasciculus.

Sensory Functions and Dysfunctions of the Central Nervous System

SOMATOSENSORY CORTEX

Postcentral Gyrus (Also Called Primary Somatosensory Area 1, or SS1)

- SS1 is responsible for the detection of incoming sensory information from the periphery.
- It is NOT responsible for the Interpretation of sensory data.
- All sensory data go through the Thalamus before reaching the Postcentral Gyrus.
- SS1 is also the site of the Sensory Homunculus.

Sensory Homunculus

- Located in the Postcentral Gyrus.
- It is the Cortical Representation of every body part's sensation.
- It is the Somatotopic Organization of Body Sensation from the Contralateral side of the body.
- Akin to the Motor Homunculus in the Precentral Gyrus.
- Face, Hands, and Mouth have a lot of representation for exploration of the external world.
- Like the Motor Homunculus, the Sensory Homunculus does NOT have a permanent organization.
- There is a plasticity to the Homunculi, should injury or disease occur.
- In people who read Braille, the reading finger has a very large representation in the primary somatosensory cortex.
- In Amputees, other body parts appropriate the cortical representation or region that had been used for the amputated part.

Lesions to the Primary Somatosensory Area (SS1)

- Loss of Sensation of the Contralateral Body Part.
- The Loss will depend on which part of the Sensory Homunculus was damaged.

Secondary Somatosensory Area 2 (SS2)

- Responsible for the Interpretation of sensory information.
- This is where meaning is attached to incoming sensory data.
- Located just posterior to the Postcentral Gyrus.

Lesions to the Secondary Somatosensory Area 2 (SS2)

- Tactile Agnosia: The umbrella term for the inability to attach meaning to sensory data.
- Tactile Agnosias Include: Astereognosis, Two-Point Discrimination Disorder, Agraphesthesia, Extinction of Simultaneous Stimulation (all described in the following sections).

Oliver Sacks Story

- Sacks describes a 60-year-old woman who was blind at birth. She was overprotected and never encouraged to explore her environment with her hands.
- As an adult, she couldn't identify objects or sensations with her hands despite having intact sensory anatomy.
- Her sensory cortices were never stimulated.
- At 60, she had to re-create the region in the cortical map for her hands by learning to use her hands to identify objects/sensations.
- This is an example of how the cortex is use-dependent.

From: The Man who Mistook His Wife For A Hat. *New York: Harper Perennial. 1987.*

The Impermanence of Cortical Representation-of Sensory and Motor Functions is true for ALL Cortical Areas—not just the Homunculi.

- In people who become deaf, regions of the Auditory Cortex are reallocated for Visual Use.
- In people who become blind, regions of the Visual Cortex are reallocated for Tactile and Auditory Use.

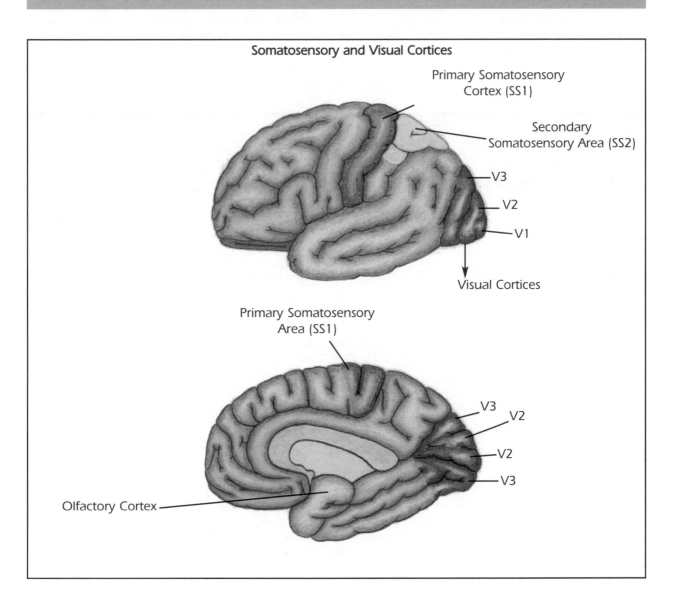

Somatosensory and Visual Cortices

Primary Somatosensory Cortex (SS1)

Secondary Somatosensory Area (SS2)

V3
V2
V1

Visual Cortices

Primary Somatosensory Area (SS1)

V3
V2
V2
V3

Olfactory Cortex

SOMATOSENSORY CORTEX: TWO CLASSIFICATIONS OF SENSATION

Primary Sensation

- Includes Pain and Temperature, Light Touch, Pressure, Vibration, and Proprioception.
- Primary Sensation is mediated by the Skin Receptors, Spinal Nerves, Spinal Tracts (Dorsal Columns and Spinothalamic Tracts), and the Somatosensory Area (SS1).

Cortical Sensation

- Cortical Sensations are mediated by the Secondary Somatosensory Cortex (SS2) and the Posterior Multimodal Association Cortex.
- Cortical Sensations include the following Tactile Agnosias:
- ASTEREOGNOSIS: The inability to identify objects by touch alone (sensory anatomy is intact; Cortical Interpretation is impaired).
- TWO-POINT DISCRIMINATION: Loss of the ability to determine whether you've been touched by one or two points (sensory anatomy is intact; the Cortical Interpretation of sensation is impaired).
- AGRAPHESTHESIA: Loss of the ability to interpret letters written on the contralateral hand of the affected side—despite intact sensory anatomy. Cortical Interpretation of sensation is impaired.
- EXTINCTION OF SIMULTANEOUS STIMULATION: Also called Double Simultaneous Extinction. The patient is touched simultaneously on two different body regions—one on the involved side; one on the uninvolved side. Extinction occurs when the patient cannot feel the tactile sensation on the involved side (even though he or she could if the tactile sensation was applied only to the involved side). Extinction of Simultaneous Stimulation occurs because the neurons that carry tactile sensation from the involved side are cortically overridden by the neurons carrying sensation from the uninvolved side.

Sensory Signs and Symptoms	Suspect
• Astereognosis, Agraphesthesia	• Contralateral Parietal Lobe Impairment
• Decreased Sensations to Pin (Hemihyperesthesia)	• SC Tract Damage
• Only the Distal Limbs present with Hyperesthesias	• Peripheral Neuropathy

VISUAL CORTEX

Primary Visual Cortex (V1)

- Receives visual information from the Optic Tracts, Lateral Geniculate Nuclei of the Thalamus, and the Superior Colliculi of the Midbrain.
- Responsible for the Detection of Visual Input—NOT Interpretation.
- Located at the most posterior region of the Occipital Lobe.

Lesions to the Primary Visual Cortex (V1)

- Loss of Sight (despite intact visual anatomy).

Lesions in V1 Produce Cortical Blindness

- When V1 is lesioned, visual information appears to travel through the Retina, Optic Nerves, Optic Tracts, and to the Lateral Geniculate Nucleus of the Thalamus. Here the visual information may be rerouted to the Visual Association Areas, rather than first proceeding to the Primary Visual Cortex (V1).
- This type of sight is known as Blind Sight or Unconscious Sight.
- When patients with V1 damage are asked to guess about observed objects in motion, or to distinguish colors, their ability to do so is better than random.
- This suggests that visual information may be rerouted from the Lateral Geniculate Nucleus to the Visual Association Areas.
- The patients, however, are not consciously aware of having seen anything and are often surprised that their guesses are so accurate. Their sight was not processed at a conscious level.

Visual Association Areas (V2 and Up)

- The Visual Cortex has more than 30 specialized areas.
- V2 (and up) is responsible for the Interpretation of Visual Input.
- This is where meaning is attached to incoming visual data.
- V1 has a mature appearance at birth. The Visual Associations Areas do not. They appear to be dependent upon the acquisition of experience.

Lesions to the Visual Association Areas

- Lesions to the Visual Associations Areas result in Visual Agnosias—an Umbrella Term that denotes the inability to attach meaning to visual data.

V2

- V2 sends visual information from V1 to the appropriate Visual Association Area for Interpretation.
- V2 processes information for further refinement by the Specialized Visual Association Areas.

V3

- V3 is responsible for interpreting Form Discrimination—the ability to recognize identifiable shapes.

Lesion to V3

- No one has ever reported a complete and exclusive loss of Form Vision.
- Area V3 forms a ring around V1 and V2. A lesion large enough to destroy all of V3 would almost certainly destroy V1, causing total blindness.
- V4 would also have to be knocked out since V4 plays a role in Form Discrimination (with regard to Line Orientation).

V4

- V4 is responsible for the Interpretation of Color Vision and Line Orientation.
- When people view an abstract color painting with no recognizable shapes, V4 has the highest Cerebral Blood Flow on PET Scans.

Lesion to V4

- A lesion to V4 produces Achromatopsia in which people only see shades of gray.
- This differs from color blindness, due to damaged Cone Receptors in the Retina.
- In Achromatopsia, people cannot recall or bring up in memory what colors look like.
- If their Retinas and V1 Regions remain intact, their knowledge of Form, Depth, and Motion are preserved.

V5

- V5 is responsible for interpreting Visual Motion (identifying objects that are in motion).
- When people view a pattern of moving black and white squares on a computer screen, V5 has the highest Cerebral Blood Flow on PET Scans.

Lesion to V5

- A lesion to V5 produces Akinetopsia in which people neither see nor understand the world in motion.
- While at rest, objects may be perfectly visible and understandable.
- Objects in motion appear to vanish.
- The other attributes of vision (color and form) remain intact.

- ALL VISUAL AREAS CONNECT DIRECTLY AND RECIPROCALLY WITH EACH OTHER.
- THE VISUAL ASSOCIATION AREAS PROJECT TO THE POSTERIOR MULTIMODAL ASSOCIATION AREA—a cortical region where somatosensory, visual, and auditory information are integrated and interpreted in relation to each other.

Oliver Sacks Story

- Sacks described a man who was legally blind since early childhood. When the man's sight was surgically restored in his 40s, he could not integrate perception of color, form, and motion into a coherent visualization that had meaning—even though his visual anatomical structures were now intact.
- This suggests that the visual association areas only develop with direct experience and use.

From: An Anthropologist on Mars. *New York: Vintage. 1995.*

In cultures located in Deep Rain Forests, Depth Perception does not develop because there are no opportunities to experience expansive views.

Sequence of Color Vision

V1   Primary visual area, sends visual information to V2 for further visual processing.

↓

V2   Sends visual color information to V4.

↓

V4   Interprets color vision. If V4 is damaged, people see the world in shades of gray.

↓

- Visual information is eventually sent to the posterior Multimodal Association Area to be integrated with other sensory data.

↓

- This information is then sent to the Hippocampus for the Storage of Visual Memories and to other areas of the Limbic System for the Integration of visual Information with Emotion.

AUDITORY CORTEX

Primary Auditory Cortex (A1)
- Located within the Insula in the Temporal Lobe.
- Responsible for Detecting sounds from the environment.
- A1 sends Auditory Information to the Auditory Association Areas for Interpretation.

Lesion to A1
- Unilateral Deafness occurs if only one hemisphere is lesioned.
- Bilateral Deafness occurs if both hemispheres are involved.

Auditory Association Areas
- There are several Auditory Association Areas located throughout the Cortex.
- They are responsible for the Interpretation of Auditory Data.
- These areas have not as yet been mapped as precisely as the Visual Associations have.
- Different Auditory Association Areas interpret specific Auditory Information: animal sounds, human language, the sounds of machinery, etc.

Lesion to the Auditory Association Areas
- A lesion to the Auditory Association Areas results in Auditory Agnosia—an Umbrella Term that denotes the inability to attach meaning to specific sounds.

Broca's Area
- Broca's Area is located ONLY in the Left Hemisphere.
- Mediates the Motoric Aspects of Speech and is responsible for the Verbal Expression of Language.
- Located just above the Lateral Fissure; sits within the Inferior Frontal Gyrus in the Premotor Area.

Lesion to Broca's Area
- Produces Expressive Aphasia or Nonfluent Speech.
- Patients can understand what is spoken to them but they cannot form meaningful sentences.

Wernicke's Area
- Also ONLY located in the Left Hemisphere.
- Sits within the Superior Temporal Gyrus in the Temporal Lobe.
- Responsible for the Comprehension of the Spoken Word.

Lesion to Wernicke's Area
- Patients cannot understand what is spoken to them.
- But they can produce intact sentences (although the meaning of their words do not relate to what was spoken to them by others).

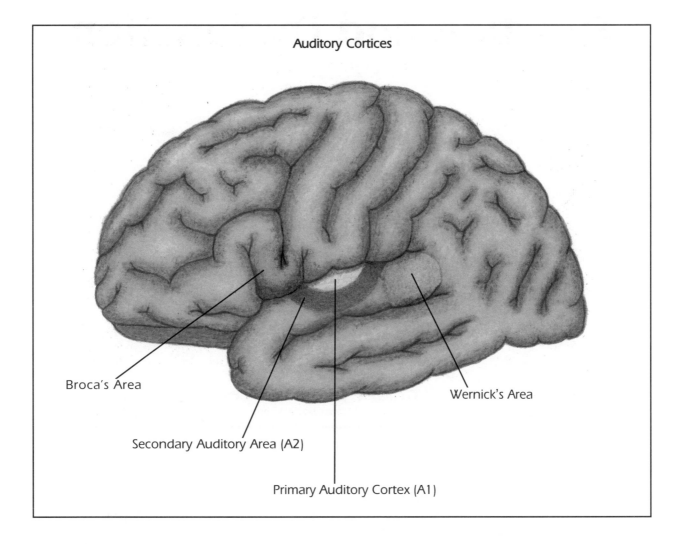

Auditory Cortices

Broca's Area

Wernick's Area

Secondary Auditory Area (A2)

Primary Auditory Cortex (A1)

Thalamus and Brainstem Sensory and Motor Roles: Function and Dysfunction

THALAMUS

- The Thalamus is part of the Diencephalon.
- The Thalamus is a Relay and Processing Center for Sensory and Motor Information.
- There are two Thalamic Lobes—one in each hemisphere.
- Contains 26 Pairs of Nuclei.
- Ventrolateral Nucleus projects to the Primary Motor Area (M1).
- Lateral Geniculate Nucleus projects to the Primary Visual Area (V1).
- Medial Geniculate Nucleus projects to the Primary Auditory Area (A1).
- Ventral Posterolateral Nucleus projects to the Primary Somatosensory Area (S1).

Thalamic Pathways

Sensory Afferent Pathway

- Sensory Receptors in the Peripheral Nervous System (PNS) project sensory messages to the Spinal Nerves. The Spinal Nerves carry sensory information to the Spinal Cord (SC) Tracts. Sensory information is then projected to the Brainstem where it is processed by the Reticular Formation. Sensory information then travels to the Thalamus, and finally to the Cortex (SS1).
- Sensory information is carried to the SC from the Spinal Nerves in the PNS. Sensory information then travels from the SC to the Brainstem where it is projected through the Middle Cerebellar Peduncle to the Cerebellum. The Cerebellum then sends the sensory information through the Superior Cerebellar Peduncle to the Thalamus. At the Thalamic level, sensory information is either rerouted to the Cortex or back down through the Brainstem.

Motor Efferent Pathway

- Motor Messages are sent from the Cortex (Primary Motor Area—M1) to the Thalamus. At the Thalamic level Motor Messages are sent down through the Brainstem, SC, and to the Muscles in the PNS. Or Motor Messages are projected from the Thalamus to the Cerebellum, SC, and to the Motor Neurons in the Ventral Horn.

Ansa Lenticularis Pathway

- The Ansa Lenticularis Pathway carries Motor Messages from the Basal Ganglia to the Ventrolateral Nucleus of the Thalamus. The information is then sent to the Primary Motor Area (M1).
- This pathway allows the Basal Ganglia to communicate with the Cortex.

Superior Colliculi Pathway

- The Superior Colliculi of the Midbrain receives sensory messages from the Optic Pathways and the Thalamic Lateral Geniculate Nucleus.
- The Superior Colliculi then sends this information to the Thalamus via the Medial Longitudinal Fasciculus.

- This pathway carries information that controls the Position of the Eyes and Head in response to Visual Information.

Inferior Colliculi Pathway

- The Inferior Colliculi of the Midbrain receives sensory information from the Thalamic Medial Geniculate Nucleus and the Auditory Cortex.
- The Inferior Colliculi then projects this sensory information back to the Thalamus and Auditory Cortex for the further Processing of Auditory Information.

Thalamic Mediodorsal Nucleus Pathway

- The Thalamic Mediodorsal Nucleus receives and projects sensory information to and from the Amygdala, Substantia Nigra, and the Temporal Cortex.
- When the Mediodorsal Nucleus is lesioned, Memory Loss occurs.

Lesions to the Thalamus

Thalamic Syndrome

- Thalamic Syndrome results from Vascular Insufficiency (eg, cerebrovascular accident).
- Thalamic Syndrome involves an alteration of Sensory Perception.
- Patients may become either Hyper- or Hyposensitive to Sensation (particularly pain and noxious stimuli) on the Contralateral Side of the Lesion.
- Several weeks after onset, patients develop burning, agonizing pain in the affected body parts.

Thalamic Tumors

- Patients present with specific symptoms depending on which Thalamic Nuclei are involved.
- A Lesion to the Ventrolateral Nucleus destroys communication with M1. Results in Paralysis of the involved body part.
- A Lesion to the Lateral Geniculate Nucleus destroys communication with V1. This can result in Cortical Blindness.
- A Lesion to the Medial Geniculate Nucleus destroys communication with A1. Cortical Deafness or Hyper/Hyposensitivity to sound can occur.
- A Lesion to the Posterolateral Nucleus destroys communication with SS1. Paresthesias, Hypoesthesia, and/or Causalgia can occur.

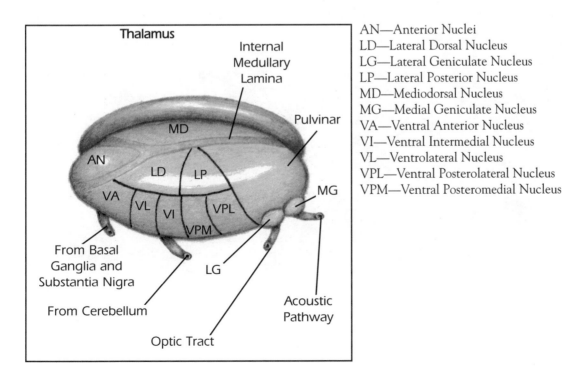

AN—Anterior Nuclei
LD—Lateral Dorsal Nucleus
LG—Lateral Geniculate Nucleus
LP—Lateral Posterior Nucleus
MD—Mediodorsal Nucleus
MG—Medial Geniculate Nucleus
VA—Ventral Anterior Nucleus
VI—Ventral Intermedial Nucleus
VL—Ventrolateral Nucleus
VPL—Ventral Posterolateral Nucleus
VPM—Ventral Posteromedial Nucleus

BRAINSTEM

- The Brainstem controls Vegetative Functions: Respiration and Primitive Stereotyped Reflexes.
- The Primitive Stereotyped Reflexes include the Cough and Gag Reflex, the Pupillary Response, and the Spontaneous Swallowing Reflex.
- The Brainstem also controls the Reticular Formation.

Reticular Formation

- The Reticular Formation is a poorly localized area of the Brainstem.
- The Reticular Activating System is thought to be located in the Midbrain.
- The Reticular Inhibiting System is believed to be located in the Pons.
- The Reticular Formation acts as a screen for incoming Sensory data to the Cortex, and outgoing Motor information from the Cortical and Subcortical Areas.

Reticular Activating System

- The Reticular Activating System alerts the Cortex to attend to important incoming sensory information.
- This process is important for filtering unnecessary data so that the individual can focus on other activities more intently.
- When important sensory data is detected by the Reticular Activating System, it alerts the Cortex—which then arouses the body and prepares it for activity.
- The Cortex uses the Thalamus, Hypothalamus, Limbic System, Vestibular System, and the Autonomic Nervous System to arouse the body.

Reticular Inhibiting System

- The Reticular Inhibiting System receives messages from the Cortical and Subcortical Centers to calm the body in response to specific sensory data.
- The Reticular Inhibiting System works conjointly with the Vestibular System to calm the body via such sensory stimulation as slow rocking, deep pressure, and vibration.

Sensory Processing Disorder of the Reticular Formation

- In children with Attention Deficit Hyperactivity Disorder (ADHD), Learning Disability (LD), Autism, and Tactile Defensiveness, the Reticular Activating System may be working in overdrive.
- These children cannot screen out extraneous sensory data. Thus they react to almost all sensory stimulation in the environment in a random disorganized fashion.
- Conversely, some children's Reticular Inhibiting System may be working in overdrive. They cannot register the sensory information that is attempting to enter the NS.
- Children with Sensory Integrative Disorders may seek out specific types of stimulation in order to obtain the type of sensory data that their NSs require to develop and function optimally.
- Sensory Integration Treatment allow the child to seek out his or her own level of sensory stimulation to help the brain process sensory information more effectively so that the child can make more adaptive responses to the environment.
- For children with Hyper-Reticular Activating Systems, Slow Rocking, Deep Pressure, and Vibration may be the types of Vestibular Sensation needed to calm the body.
- For children with Hyper-Reticular Inhibiting Systems, sensory stimulation that arouses the brain is likely needed.

Pharmacologic Treatment: RITALIN
- Ritalin is a drug that attempts to balance the divisions of the Reticular Formation.
- Ritalin can have the affect of calming down a Reticular Activating System that is in overdrive.
- It may also stimulate the Reticular Activating System in a child whose Reticular Inhibiting System is dominant.

Brainstem Damage and Coma

- Severe Brainstem Damage can result in Coma.
- There are two basic types of Comatose States: Persistent Vegetative State and Brain Death.

Persistent Vegetative State
- In a Persistent Vegetative State the Brainstem—including the Reticular Activating System—remains intact.
- The brunt of neurologic destruction is located in the Cerebral Hemispheres.
- A Persistent Vegetative State often results from Cardiac or Respiratory Arrest causing a lack of blood flow (Ischemia) and the loss of oxygen (Hypoxia) to the brain for several minutes.
- The Brainstem is fairly resistant to Ischemia and Hypoxia. 4 to 6 minutes of complete blood and oxygen loss to the brain can result in extensive cortical destruction while sparing the Brainstem.
- After a period of days to a month, the patient will awaken into a condition of eyes-opened unconsciousness. This is called a Vegetative State.
- Because the Brainstem is spared, the Cough, Gag, and Swallowing Reflexes remain intact.
- This decreases the likelihood of Respiratory Infections and substantially lengthens life expectancy (particularly with life support technological systems).

Brain Death
- Brain Death occurs when all Brainstem functions are lost.
- The Heartbeat can continue because it is semi-autonomous from Autonomic Nervous System regulation.
- This type of Coma is a state of sleep-like (eyes-closed) unarousability due to extensive damage to the Reticular Activating System. The Cough, Gag, and Swallowing Reflexes are lost, leading to Fatal Respiratory Infections in 6 months to a year.

Cranial Nerve Nuclei Damage in the Brainstem
- The Cranial Nerve Nuclei are located in the Brainstem.
- They innervate ipsilateral body structures.
- Brainstem disorders depend upon which Cranial Nerve Nuclei have been lost.

Spinal Cord Tract Damage in the Brainstem
- Descending Motor Tracts and Ascending Sensory Tracts travel through the Brainstem and mediate motor and sensory functions of structures on the Contralateral side of the body.
- When Spinal Cord tracts are lost, due to Brainstem damage, Hemiplegia and Hemiparesthesias on the Contralateral side of the body result.

Right Vs. Left Brain Functions and Disorders

HEMISPHERAL DOMINANCE

- Generally one side of the Brain has Dominant Control.
- Specific individuals may be Left or Right Brain Dominant.
- In most people, the Left Hemisphere is Dominant.
- Hemispheral Dominance may be reversed in some people who are Left Hand Dominant. Or only certain Brain Functions may be dominant in the Right Hemisphere.

LEFT BRAIN FUNCTIONS

- The Left Hemisphere controls Motor Function on the Right Side of the Body.
- The Left Hemisphere receives Sensory Information from the Right Side of the Body.
- The Left Hemisphere has a role in Language—specifically the Interpretation and Expression of Aural and Written Words. The Left Hemisphere specifically interprets the Concrete Meanings of Words (as opposed to the Abstract or Symbolic Meaning of Words).
- For example, the Left Hemisphere interprets the Literal Meaning of a Story—but NOT the hidden or symbolic meaning.
- The Left Hemisphere controls Concrete Functions that can be easily observed and measured:
  - Interpreting the Concrete Meaning of written or spoken words
  - Math Calculations
  - Writing the Letters of the Alphabet
  - Reading a Sentence
  - Categorizing Shapes
  - Sequencing Steps in a Task

RIGHT BRAIN FUNCTIONS

- The Right Hemisphere Controls Motor Function on the Left side of the body.
- The Right Hemisphere receives Sensory Information from the Left side of the body.
- The Right Hemisphere has a role in the Interpretation of Perception—how humans perceive their environment.
- The Right Hemisphere also has a role in the Interpretation of information that is Abstract and Creative (as opposed to Concrete and Logical).
- The Right Hemisphere controls Abstract Functions that cannot be easily observed—functions that relate to the Perception of oneself in relation to the environment.
- The Right Hemisphere also has a role in Language, but it is the Interpretation of the Abstract or Symbolic Meaning of a Story or Joke. Or interpreting someone's Verbal Tone and Gestures.

- Perception includes: Visual and Spatial Perception, Language Perception, Motor Planning Perception, Body Schema Perception, and Tactile Perception.

LEFT HEMISPHERE DISORDERS

- Wernicke's and Broca's Aphasia.
- Contralateral Motor and Sensory Problems.
- Acalculia: the inability to calculate math problems.
- Agraphia: the inability to write words (that the patient was once able to pre-injury/disease).
- Alexia: the inability to read written words (that the patient was once able to pre-injury/disease).
- Contralateral Motor and Sensory Problems.

RIGHT HEMISPHERE DISORDERS

- Right Hemisphere Disorders involve an impairment in the Recognition of Physical Reality.
- Right Hemisphere Disorders distort Physical Reality. They distort the environment and/or one's own body perception.
- Visual-Spatial Disorders.
- Body Schema Perception Disorders.
- Apraxias: Motor Planning Perceptual Problems.
- Perceptual Language Disorders.
- Contralateral Motor and Sensory Problems.

THE CORRELATION BETWEEN ANATOMICAL DAMAGE AND SYMPTOMATOLOGY

- It is difficult to correlate precise anatomical damage with specific symptomatology.
- This is due to Individual differences in Human Brains.
- Each Human Brain varies with regard to Sulci and Gyri Patterns.
- It is also difficult to correlate damage and symptomatology because of Neuroplasticity.

NEUROPLASTICITY

- Neuroplasticity occurs when other areas of the brain assume the functions once mediated by regions that have been damaged.
- Human Brains appear to possess a vast amount of brain matter that does not become active until damage occurs to other areas. At this time, those previously unused regions become active and take over the function of the damaged areas.
- It is possible that the same brain function may be shared by several, separate brain regions that lie dormant until injury/disease occurs. This may be the brain's own evolutionary attempt at compensation.
- Neuroplasticity is most viable in children because the central nervous system is not fully matured, but rather is still developing.

Neuroplasticity Is Dependent Upon

- The Severity of the Neurologic Damage.
- The Individual's Age.
- Premorbid Health Status.
- Pre-injury use of the Damaged Brain Areas—the more an individual has used a particular brain function premorbidly, the more capacity one has of resolving that damaged brain area to some degree.

Perceptual Functions and Dysfunctions of the Central Nervous System

MULTIMODAL AREAS OF THE CENTRAL NERVOUS SYSTEM

- The Multimodal Areas of the CNS are also called Convergence Association Areas.
- This is where sensory information that has been processed and interpreted by the Somatosensory Association Area, the Auditory Association Areas, the Visual Association Areas, and the Motor Association Area (Prefrontal Cortex) all converge to allow for the further integration of all types of sensory data.
- There are two Multimodal Association Areas in the Brain:
  - The Posterior Multimodal Association Area
  - The Anterior Multimodal Association Area (the Prefrontal Cortex)

Posterior Multimodal Association Area

- This Multimodal Association Area is located in the posterior of the brain in a region where the Parietal, Occipital, and Temporal Lobes meet.
- The Posterior Multimodal Association Area integrates sensory information processed by the Somatosensory Association Areas, the Visual Association Areas, and the Auditory Association Areas.
- This is where sound is connected with visual and tactile information.
- For example, the Posterior Multimodal Association Area connects the sound of the ocean with the color of the waves and crests, and the feel of the sand on one's feet.
- After integrating the three types of sensory data, the Posterior Multimodal Association Area then sends this information to the Anterior Multimodal Association area (the Prefrontal Cortex).

Anterior Multimodal Association Area

- The Anterior Multimodal Association Area is located in the Prefrontal Cortex—one storage area for Motor Plans.
- The Anterior Multimodal Association Area takes the integrated sensory data from the Posterior Multimodal Association Area and uses it to make decisions about which Motor Plan to Implement.
- Once a decision is made, the Anterior Multimodal Association Area sends this information to the Premotor Area to Access the Appropriate Motor Plan.
- Once the Appropriate Motor Plan is Accessed, the Primary Motor Area (M1) implements the Motor Plan.
- For example, one may decide to implement the Motor Plans required for swimming in the ocean.

Impairment of the Multimodal Association Areas

- Because the Multimodal Association Areas in the Right Hemisphere play an important role in Perception (how one perceives the environment and one's relationship to the environment), damage to the Multimodal Association Areas often result in Perceptual Disorders.

Perceptual Impairment

- Perceptual Impairment more often involves dysfunction of the Right Hemisphere (specifically the Posterior Multimodal Association Area), rather than the Left.
- Right Hemisphere disorders of the Posterior Multimodal Association Area involve an impairment in the Recognition of Physical Reality. Physical Reality has become distorted.
- One's own relation to the environment is distorted. One's own relation to one's body is also distorted.
- There are several classifications of Perception:
  - Visual Perception (including Spatial Perception)
  - Language Perception (including Expressive and Receptive Language Perception)
  - Body Schema Perception
  - Motor Planning Perception (or Praxis)
  - Tactile Perception

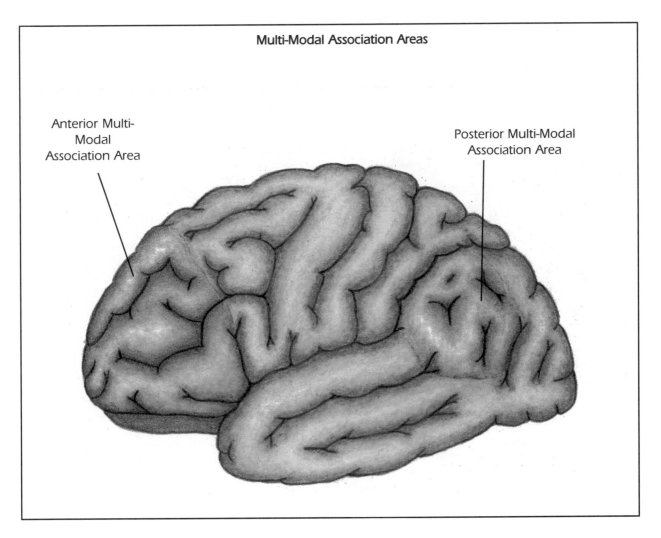

Multi-Modal Association Areas

Anterior Multi-Modal Association Area

Posterior Multi-Modal Association Area

VISUAL PERCEPTUAL DISORDERS

Visual Agnosia

- Visual Agnosia is an Umbrella Term for the inability to identify and recognize familiar objects and people (although the visual anatomy is intact).
- Lesions that cause Visual Agnosias are usually located in the Right Hemisphere in the Posterior Multimodal Association Area.

Prosopagnosia

- Prosopagnosia is the inability to identify familiar faces because the individual cannot perceive the unique expressions of facial muscles that make each human face different from each other.
- The Left Hemisphere specializes in identifying faces through facial details. So the individual may be able to identify his wife because she has a mole on her nose. However, this form of pattern recognition causes the individual to think that everyone with a mole on her nose is his wife.

Simultanognosia

- Simultanognosia involves difficulty interpreting a visual stimulus as a whole.
- Patients often Confabulate to Compensate for what they cannot interpret visually.

Oliver Sacks Story

Prosopagnosia

The patient was a young man of 32 who, following a severe car accident, was unconscious for 3 weeks. He complained of an inability to recognize faces, even those of his wife and children. Not a single face was familiar to him but there were three he could identify—these were work colleagues: one with an eye blinking tic, one with a large mole on his cheek and a third who was extremely tall and thin (no one else was like him). Each of these men was recognizable solely by the single prominent feature mentioned. In general he recognized familiar people only by their voices. He had difficulty even recognizing himself in the mirror. (p.21)

Simultanognosia

I opened up a copy of *National Geographic* and asked the patient to describe some pictures in it. His responses here were very curious. His eyes would dart from one thing to another, picking up tiny features, individual features, as he had done with my face when identifying me. A striking brightness, a color, a shape would arrest his attention and elicit comments. But in no case did he get the scene as a whole. He failed to see the whole, seeing only details, which he spotted like blips on a radar screen. He never entered into relation with the picture as a whole. He had no sense whatsoever of a landscape or a scene. I showed him the cover—an unbroken expanse of Sahara Dunes. "I see a river," he said. "And a little guest house with its parasols here and there." He was looking right off the cover into mid-air and confabulating nonexistent features. (p.10-11)

From: The Man Who Mistook His Wife For A Hat. *New York: Harper Perennial. 1987.*

Metamorphosia

- Metamorphosia involves a visual distortion of the physical properties of objects, so that objects appear bigger, smaller, or heavier than they really are.

Color Agnosia

- In Color Agnosia Patients appear to forget the concept of color.
- They do not appear to know the color of common objects.
- For example, a patient might think that a banana was blue.

Color Anomie

- Patients have lost the names for colors.
- But they would continue to recognize that a blue banana was strange.

COLOR AGNOSIA and COLOR ANOMIE differ from CEREBRAL ACHROMATOPSIA—a condition that occurs when V4 is lost and the world appears in shades of gray. Cerebral Achromatopsia erases the memory of color.

VISUAL-SPATIAL PERCEPTUAL DYSFUNCTION

Right-Left Discrimination Dysfunction
- Involves difficulty understanding and using the concepts of right and left.
- Quick Screen:
    - Have the patient point to right and left body parts.
    - Give patients left and right directions around the treatment facility.

Figure-Ground Discrimination Dysfunction
- Involves difficulty distinguishing the foreground from the background.
- Quick Screen:
    - Have the patient pick out forks from a drawer of disorganized utensils.
    - On a meal plate, observe whether a patient can distinguish the potatoes from the background of the white dish.

Form-Constancy Dysfunction
- Involves difficulty attending to subtle variations in form, or changes in form such as a size variation of the same object.
- Quick Screen:
    - Can the patient still identify an object when it is turned on its side or placed upside down?
    - Can a patient identify a group of shapes as triangles when they are different sizes?

Position in Space Dysfunction
- Involves difficulty with concepts relating to positions such as up/down, in/out, behind/in front of, before/after.
- Quick Screen: Give the patient directions using the above terms:
    - Place the pencil on top of the box and place the box inside the drawer.
    - Take the pot from underneath the sink and place it on the counter. Put the rice inside the pot.

Topographical Disorientation
- Involves difficulty comprehending the relationship of one location to another.
- Quick Screen:
    - Can the patient find his way around the hospital using written directions or a pictorial map?

Depth Perception Dysfunction (Stereopsis)
- Involves difficulty determining whether one object is closer to the patient than another object.
- Quick Screen:
    - Can the patient determine whether objects in the natural environment are near or far in relation to each other and in relation to the patient?

BODY SCHEMA PERCEPTUAL DYSFUNCTION

- Body Schema is the awareness of spatial characteristics of one's own body—an awareness formed by current and previous sensory input.
- It is the Neural Perception of one's body in space.
- This neural perception is derived from a synthesis of tactile, proprioceptive, and pressure sensory associations about the body and its parts.
- Body Schema Perceptual Dysfunction is more likely to occur as a result of Right Hemisphere Lesions in the Posterior Multimodal Association Area.
- However, as with all of the Perceptual Disorders, they can also result from Left Hemisphere Lesions.

Body Schema Differs from Body Image
- Body Image is the emotional and cognitive assessment one holds about one's body.
- For example, a patient with anorexia may experience her Body Image as fat and ugly. However, with regard to her Body Schema, she may be overly aware of sensory input from stomach distension after a meal. Someone without anorexia may not be as consciously aware of stomach distension.

- Patients with Anosagnosia (severe neglect) may have positive thoughts about their Body Image, believing that it is just fine. With regard to their Body Schema, they have no awareness that the left side of their body exists— they are not processing the neural input from that body part.

Finger Agnosia

- Finger Agnosia involves an impaired perception concerning the relationship of the fingers to each other.
- Also impaired identification and localization of one's own fingers.
- Quick Screen: Ask the patient to tap her index finger or to touch her ring finger with her thumb.

Unilateral Neglect

- Involves the inability to integrate and use perceptions from the left side of the body or the left side of the environment.
- The awareness of the left side of environment is lost temporarily.
- Patients more often have a Left Neglect rather than a Right Neglect; although Right Neglect Syndromes occur as well.
- A Left Neglect often results from a Lesion to the Right Hemisphere's Posterior Multimodal Association Area.
- Patients with Unilateral Neglect can be trained to heighten their awareness of the left side of their bodies and environments.
- Quick Screen:
  - Ask the patient to draw a Clock, a Human Figure, Read a Paragraph, Perform a Cross-out the Hs Exercise.
  - Observe the patient while eating a meal. Check whether the patient attends to both sides of the plate or ignores one half.

Anosagnosia

- Anosagnosia is extensive neglect and failure to recognize one's own body paralysis on one side.
- While patients with Unilateral Neglect can usually be taught to enhance their awareness of the left side of the body and environment, patients with Anosagnosia cannot be taught in the same way.
- Anosagnosia is accompanied by a strange Affective Dissociation. Patients show extraordinary indifference to the affected limb.
- They have no concept that they have a paralyzed limb.
- Patients may ask the hospital staff to "take the arm away with the lunch tray."
- Anosagnosia is usually a transient state of the acute CVA patient. Anosagnosia will usually resolve as the patient recovers.
- Right Hemisphere Damage is considered to cause more severe Neglect Syndromes than do Left Hemisphere Lesions.
- Right Hemisphere Damage that produces Left Neglect resolves less readily than Right Neglect Syndromes.
- Patients who are several months post neurologic damage are more likely to display Left Neglect because patients with Right Neglect have usually cleared by that stage.
- Sometimes patients with severe Left Neglect often have to deal with the Neglect Syndrome for the remainder of their lives—especially if the patient is elderly or the neurologic damage was severe.

Extinction of Simultaneous Stimulation (Double Simultaneous Extinction)

- Extinction of Simultaneous Stimulation is categorized as a Cortical Somatosensory Association Disorder.
- However, it is also considered to be a form of Attentional Neglect.
- The therapist touches the patient on 2 regions: (a) on the involved side, and (b) on the uninvolved side.
- Extinction occurs when the patient cannot feel the tactile sensation on the involved side (even though he could if the tactile sensation was only applied to the involved side).
- As the neurons mediating tactile sensation recover on the involved side, they can be overridden by tactile stimulation on the uninvolved side.
- This is called a Limited Attention Recovery Phase.

LANGUAGE PERCEPTION DYSFUNCTION

The Aphasias

- Aphasia means impairment in the Expression and/or the Comprehension of Language.
- There are two Classifications of Aphasia: Receptive and Expressive.

Receptive Aphasia

- Receptive Aphasia is impairment in the Comprehension of Language.
- WERNICKE'S APHASIA is difficulty comprehending the Literal Interpretation of Language. Wernicke's Aphasia always results from a Left Hemisphere Lesion in the brain region referred to as Wernicke's Area.
- ALEXIA is the inability to comprehend the written word; also the inability to read. Alexia can occur as a result of Lesions to either Hemisphere; more often the Left Hemisphere is Lesioned.
- DYSLEXIA is the impaired ability to read. It is a language problem in which the ability to break down words into their most basic units—phonemes—is impaired.
- ASYMBOLIA is difficulty comprehending Gestures and Symbols. Usually the Left Hemisphere is Lesioned.
- APROSODIA is impaired comprehension of Tonal Inflections used in conversation. Patients have difficulty perceiving the Emotional Tone of someone's conversation. Aprosodia usually results from a Right Hemisphere Lesion.

Receptive Aphasia in the Right vs Left Hemisphere

- A Receptive Aphasic with a Left Hemisphere Lesion can still perceive and accurately interpret the Emotional Tones of a conversation, even though they cannot understand the concrete meaning of the words.
- A Receptive Aphasic with a Right Hemisphere Lesion can understand the concrete meaning of the words but not the Emotional Tone of Conversations.

Expressive Aphasia

- An Expressive Aphasia involves difficulty expressing clear, meaningful language.
- BROCA'S APHASIA is an Expressive Language Disorder in which patients can understand what is spoken to them, but they cannot express their ideas in an understandable way. Often they speak gibberish or sentences that do not make sense. Broca's Aphasia always results from a Left Hemisphere Lesion in the brain region referred to as Broca's Area.
- ANOMIE is the inability to Remember and Express the Names of People and Objects. The individual may know the person but cannot remember his name. Anomie differs from Prosopagnosia; in Prosopagnosia individuals do not recognize familiar faces. Lesions resulting in Anomie can occur in either hemisphere.
- AGROMMATION is the inability to Arrange Words Sequentially so that they Form Intelligible Sentences. Occurs as a result of Left Hemisphere Lesions.
- AGRAPHIA is the inability to Write Intelligible Words and Sentences. Occurs as a result of Left Hemisphere Lesions.
- ACALCULIA is the inability to Calculate Mathematical Problems. Dyscalculia is difficulty calculating math. Lesions that result in Acalculia occur in the Left Hemisphere.
- ALEXITHYMIA is the inability to express one's emotions through words. Dyslexithemia involves difficulty attaching word to feelings. Can occur as a result of either Left or Right Hemisphere damage.

Differences in Male and Female Language Processing

- Females use both hemispheres in the processing of language. They use the Left Hemisphere to interpret the Concrete Meaning of words and sentences, and the Right Hemisphere to interpret the Emotions attached to those words and sentences.
- Females are also more able to attach words to their emotions. This requires the ability to use the Left Hemisphere to attach words to Emotions generated in the Right Hemisphere.
- Males predominantly use the Left Hemisphere to process language. Because they do not as readily integrate both hemispheres in language processing, they may have more difficulty attaching words to their emotions, referred to as Dyslexithymia.

PERCEPTUAL MOTOR DYSFUNCTION

- Perceptual Motor Dysfunction involves the Apraxias or Motor Planning Impairments.
- Apraxia can result from either Right or Left Hemisphere Lesions—but usually result from Right Hemisphere Lesions of the Anterior Multimodal Cortex, the Premotor Area, and/or the Primary Motor Cortex.
- Patients with Apraxia have a distorted perception of the motor strategies required to negotiate their environment.
- There are several Classifications of Apraxia:
  - Ideational Apraxia
  - Ideomotor Apraxia
  - Dressing Apraxia
  - Two- and Three-Dimensional Constructional Apraxia

Ideational Apraxia

- Involves an inability to cognitively understand the motor demands of the task.
- For example, a patient may not understand that a shirt is an article of clothing to be worn on the torso and upper extremities.

Ideomotor Apraxia

- Involves the Loss of the Kinesthetic Memory of Motor Patterns—essentially the Motor Plan for a specific task is lost.
- Or the Motor Plan may be intact, but the patient cannot access the appropriate Motor Plan or may implement an inappropriate Motor Plan for a specific task.
- Such patients can cognitively understand the motor demands of the task but cannot translate that understanding into appropriate motor movements.

Dressing Apraxia

- Involves an inability to dress oneself due to an impairment in either (a) Body Schema Perception or (b) Perceptual Motor Functions.
- Example: Patients may attempt to put their arms through pant legs or dress only one half of their body.

Two- and Three-Dimensional Constructional Apraxia

- Involves an inability to copy Two- and Three-Dimensional Designs or Models.
- A patient who is an architect may be unable to draw Two-Dimensiaonl Blueprints.
- A patient who is a building contractor may be unable to put together a bird house wood kit.
- Patients with Constructional Apraxia due to a Right Hemisphere Lesion draw objects or put together pieces of a kit in a spatially disorganized way.
- Patients with Constructional Apraxia due to a Left Hemisphere Lesion draw objects that lack detail. Three-Dimensional objects are correctly spatially organized but pieces are often left out.

TACTILE PERCEPTUAL DYSFUNCTION

Tactile Agnosia

- Tactile Agnosia is the umbrella term for the inability to attach meaning to somatosensory data.
- Tactile Agnosia commonly results from lesions to the Secondary Somatosensory Area (SS2).
- One's touch and pain/temperature receptor anatomy remain intact.
- The ability to attach meaning to somatosensory data is referred to as Cortical Sensation (as opposed to Primary Sensation).

Cortical Sensation

- Cortical Sensation includes the following forms of Tactile Agnosia:
  - Astereognosis: The inability to identify objects by touch alone. Asterognosis can be further broken down into Ahylognosia and Amorphagnosia.
  - Ahylognosia: Inability to discriminate between different types of materials by touch alone.
  - Amorphagnosia: Inability to discriminate between different forms by touch alone.
  - Two-Point Discrimination: Loss of the ability to determine whether one has been touched by one or two points. An aesthesiometer is the instrument used to assess two-point discrimination.
  - Agraphesthesia: Loss of the ability to interpret letters written on the contralateral hand.
  - Extinction of Simultaneous Stimulation: The inability to determine that one has been touched on both the involved side and the uninvolved side—the neural sensation of the uninvolved side overrides the ability to perceive touch on the involved side.

Blood Supply of the Brain: Cerebrovascular Disorders

MAJOR ARTERIES

Internal Carotids (2)

Route

- The Internal Carotids rise from the Common Carotid Artery and enter the brain at the Level of the Optic Chiasm.

Supplies

- They are the Major Arteries that supply the Brain.

Vertebral Arteries (2)

Route

- The Vertebral Arteries run along the Lateral Aspect of the Medulla.
- They Connect to form the Basilar Artery at the base of the Pons-Medulla Junction.
- They give rise to the Anterior Spinal Artery.

Supplies

- They supply the Lateral Medulla Areas.

Anterior Spinal Artery (1)

Route

- The Anterior Spinal Arteries begin as two small branches that become One Main Artery.
- The two Anterior Spinal Branches rise off of the Vertebral Arteries and become One Main Artery that travels along the Anterior Surface of the Medulla and Spinal Cord.

Supplies

- The Spinal Artery supplies the Anterior Portion of the Medulla and Spinal Cord.

THREE ARTERIES THAT SUPPLY THE CEREBELLUM

Posterior Inferior Cerebellar Arteries (2)

Route

- The Posterior Inferior Cerebellar Arteries rise from the Vertebral Arteries at the Medulla Level.

Supplies

- They supply part of the Dorsolateral Medulla (including the Cerebellar Peduncles), the Inferior Surface of the Cerebellum, and the Deep Cerebellar Nuclei.

Anterior Inferior Cerebellar Arteries (2)

Route

- These rise from the Vertebral Arteries at the Pons-Medulla Junction.

Supplies
- Supplies the Inferior Surface of the Cerebellum and the Deep Cerebellar Nuclei.

Superior Cerebellar Arteries (2)

Route
- The Superior Cerebellar Arteries rise from the Basilar Artery at the Pons-Midbrain Junction.

Supplies
- Supplies the Superior Aspect of the Cerebellum and parts of the Deep Cerebellar Nuclei.

The BASILAR ARTERY Does NOT Supply the Cerebellum. But it Does Give Rise to the Superior Cerebellar Arteries.

Basilar Artery (1)

Route
- Travels along the Anterior Aspect of the Pons.
- Gives rise to the Superior Cerebellar Arteries.

Supplies
- Supplies the Anterior and Lateral Aspects of the Pons.

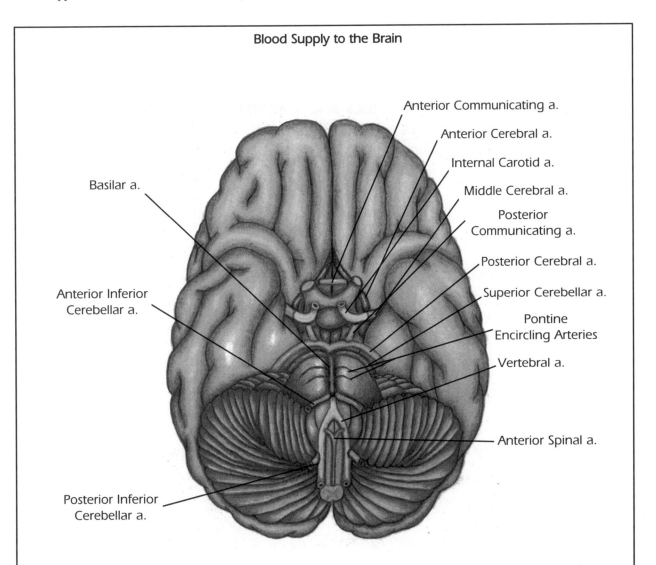

Blood Supply to the Brain

THREE MAIN CEREBRAL ARTERIES

Posterior Cerebral Arteries (2)

Route

- The Posterior Cerebral Arteries rise from the Basilar Artery.

Supplies

- Supplies the Medial and Inferior Surfaces of the Temporal and Occipital Lobes, the Thalamus, and the Hypothalamus.

Middle Cerebral Arteries (2)

Route

- Rises from the Internal Carotids and travels through the Lateral Fissure to the Brain's Surface.

Supplies

- The Middle Cerebral Arteries supply the Lateral Surfaces of the Frontal, Temporal, and Parietal Lobes.
- They also supply the Inferior Surface of part of the Frontal and Temporal Lobes.

Anterior Cerebral Arteries (2)

Route

- Rises from the Internal Carotids.

Supplies

- Supplies the Superior, Lateral, and Medial Aspects of the Frontal and Parietal Lobes.
- Also supplies part of the Basal Ganglia and the Corpus Callosum.

TWO COMMUNICATING ARTERIES AND MULTIPLE ENCIRCLING ARTERIES

- The two Communicating Arteries provide blood supply pathways to the Major Cerebral Arteries.
- The Multiple Pontine Encircling Arteries provide a blood supply pathway to the Pons.

Posterior Communicating Arteries (2)

Route

- Connect the Internal Carotids and the Posterior Cerebral Arteries.

Supplies

- Supplies the Diencephalon, Internal Capsule, and the Optic Chiasm.

Anterior Communicating Artery (1)

- The Anterior Communicating Artery Connects the two Anterior Cerebral Arteries.

Pontine Encircling Arteries (Multiple)

Route

- Rise from the Basilar Artery and wrap around the Pons.

Supplies

- They supply the Lateral and Posterior Portions of the Pons.

CIRCLE OF WILLIS

- The Circle of Willis is a Circuit of Interconnecting Arteries that Function to Prevent Lack of Blood Flow to the Brain due to Occlusion.
- Components of the Circle of Willis Include:
  - Posterior Cerebral Arteries
  - Posterior Communicating Arteries
  - Internal Carotids
  - Anterior Cerebral Arteries
  - Anterior Communicating Artery

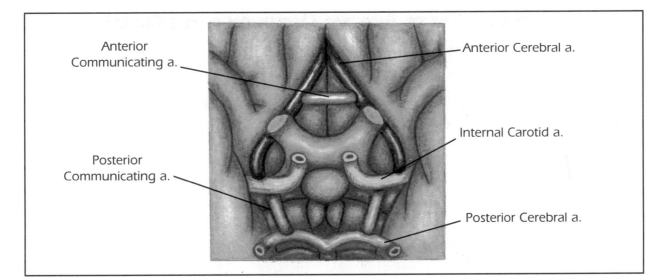

Anterior Communicating a.

Anterior Cerebral a.

Internal Carotid a.

Posterior Communicating a.

Posterior Cerebral a.

Blood Supply of Specific Brain Regions

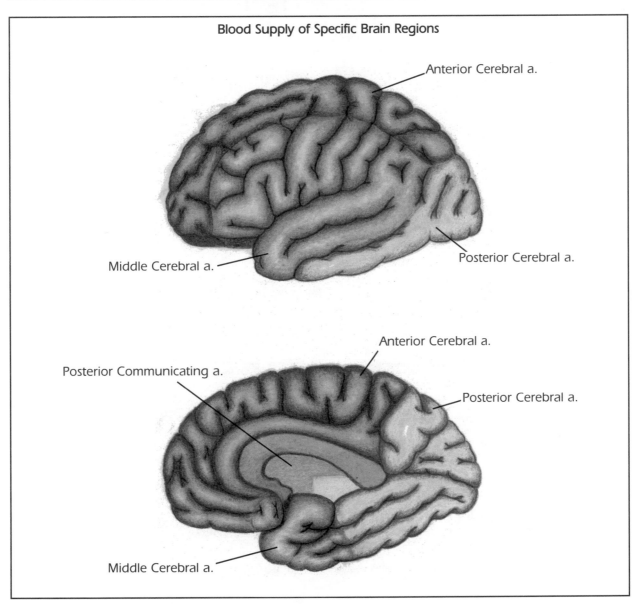

Anterior Cerebral a.

Middle Cerebral a.

Posterior Cerebral a.

Anterior Cerebral a.

Posterior Communicating a.

Posterior Cerebral a.

Middle Cerebral a.

COMMON AREAS OF ARTERIAL OCCLUSION IN THE CORTEX

Middle Cerebral Arterial Occlusion

- The Middle Cerebral Arteries are the Most Common Site of Occlusion resulting in Cerebrovascular Accident (CVA).
- The Middle Cerebral Arteries supply the Lateral Surfaces of the Frontal, Temporal, and Parietal Lobes, and the Inferior Surface of portions of the Frontal and Temporal Lobes.

Middle Cerebral Arterial Occlusion in the Left Hemisphere

- An Occlusion in the Middle Cerebral Artery in the Left Hemisphere may result in:
    - CONTRALATERAL HEMIPLEGIA: on the right side of the body. The Primary Motor Area is Lesioned.
    - CONTRALATERAL HEMIPARESTHESIA: on the right side of the body. The Primary Somatosensory Area is Lesioned.
    - APHASIA: Broca's or Wernicke's Area may be Lesioned. So may other Language Areas resulting in different types of Aphasia.
    - COGNITIVE INVOLVEMENT: Impairment in Cognitive Function results from a Frontal Lobe Lesion.
    - AFFECTIVE INVOLVEMENT: Often when the Left Hemisphere is Lesioned, the patient may display emotional lability and depression. This is sometimes referred to as a catastrophic response.

Middle Cerebral Arterial Occlusion in the Right Hemisphere

- An Occlusion in the Middle Cerebral Artery in the Right Hemisphere may result in:
    - CONTRALATERAL HEMIPLEGIA: on the left side of the body. The Primary Motor Area is Lesioned.
    - CONTRALATERAL HEMIPARESTHESIA: on the left side of the body. The Primary Somatosensory Area is Lesioned.
    - PERCEPTUAL DEFICITS: Left Neglect Syndromes are common with damage to the Right Hemisphere—particularly to the Posterior Multimodal Association Area.
    - APRAXIA: The Anterior Multimodal Association Area, Premotor Area, and/or the Primary Motor Cortex may be Lesioned.
    - COGNITIVE INVOLVEMENT: Impairment in Cognitive Function results from a Frontal Lobe Lesion.
    - AFFECTIVE INVOLVEMENT: Often when the Right Hemisphere is Lesioned, the patient may display euphoria or report a sense of well-being. If a neglect Syndrome is present, the patient is often unaware of his or her deficits.

Posterior Cerebral Arterial Occlusion

- The Posterior Cerebral Arteries supply the Medial and Inferior Surfaces of the Temporal and Occipital Lobes.
- These Arteries also help to supply the Thalamus and the Hypothalamus—however, a lesion to a Posterior Cerebral Artery will not likely affect the Thalamic and Hypothalamic Functions.
- A Lesion to one of the Posterior Cerebral Arteries may result in:
    - MEMORY LOSS (due to Temporal Lobe Involvement)
    - VISUAL PERCEPTUAL DEFICITS: result from damage of the Occipital Lobe and the Posterior Multimodal Association Area.
    - VISUAL FIELD CUTS: result from occlusion to the Optic Chiasm. The Optic Chiasm is supplied by the Posterior Communicating Arteries, which connect to the Posterior Cerebral Arteries.

Anterior Cerebral Arterial Occlusion

- The Anterior Cerebral Arteries supply the Superior, Lateral, and Medial Aspects of the Frontal and Parietal Lobes.
- These Arteries also help to supply portions of the Basal Ganglia and Corpus Callosum.
- A Lesion to one of the Anterior Cerebral Arteries may result in:
    - CONTRALATERAL HEMIPLEGIA: The Primary Motor Area is Lesioned.
    - CONTRALATERAL HEMIPARESTHESIA: The Primary Somatosensory Area is Lesioned.
    - COGNITIVE INVOLVEMENT: Due to Frontal Lobe Involvement.
    - APRAXIA: The Anterior Multimodal Association Area, Premotor Area, and/or the Primary Motor Area may be lesioned.

- AFFECTIVE INVOLVEMENT: If the Left Hemisphere is Lesioned, emotional lability and depression may occur. If the Right Hemisphere is Lesioned, euphoria or emotional dissociation may occur.

CEREBELLAR ARTERIAL OCCLUSION

- The Three Major Symptoms of Cerebellar Disorders Include:
  - Incoordination, Ataxia, and Intention Tremors

Posterior Inferior Cerebellar Arterial Occlusion

- The Posterior Inferior Cerebellar Arteries supply the Cerebellar Peduncles.
- Because the Posterior and Anterior Spinocerebellar Tracts travel through the Superior Cerebellar Peduncle, damage to the Superior Cerebellar Peduncles may result in IPSILATERAL HYPERTONICITY and HYPERACTIVE REFLEXES.
- Because the Posterior Inferior Cerebellar Arteries also supply blood to the Medulla, an occlusion to this artery may also result in VERTIGO, NAUSEA, and NYSTAGMUS as a result of Vestibular Nerve Nuclei Loss.
- Cerebellar Arterial Occlusion often involves the Brainstem Structures that are supplied by the Cerebellar Arteries.

Anterior Inferior Cerebellar and Superior Cerebellar Arterial Occlusion

- Occlusion to either of these two arteries may result in:
  - IPSILATERAL HYPOTONICITY and HYPOREFLEXIA
  - DYSMETRIA
  - ADIADOCHOKINESIA (and Dysdiadochokinesia)
  - REBOUND PHENOMENON
  - STACCATO VOICE
  - ATAXIC GAIT
  - INTENTION TREMORS
  - INCORRDINATION
- Occlusion to these Arteries may also result in the following due to the Arteries' connection to the blood supply of the Medulla:
  - VESTIBULAR SIGNS, FACIAL SENSORY IMPAIRMENT, DYSPHAGIA, DYSARTHRIA, and BELL'S PALSY

OCCLUSION OF ARTERIES THAT SUPPLY THE BRAINSTEM

Anterior Spinal Artery Occlusion

- The Anterior Spinal Arteries supply the Medulla including the Pyramids and the Vestibular, Hypoglossal, Glossopharyngeal, and Vagal Nerve Nuclei.
- CONTRALATERAL HEMIPLEGIA can result if the Pyramids are lost.
- DEVIATION OF THE TONGUE TO THE AFFECTED SIDE can result if the Hypoglossal Nerve Nuclei are lost.
- DYSPHAGIA and LOSS OF THE GAG REFLEX can result if the Glossopharyngeal and/or Vagal Nerve Nuclei are lost.
- NYSTAGMUS and BALANCE DISTURBANCES can result if the Vestibular Nuclei are lost.

Vertebral Arterial Occlusion

- The Vertebral Arteries supply the Lateral Aspect of the Low Medulla including the Accessory Nuclei.
- DYSPHAGIA may occur if the Accessory Nerve Nuclei are lost.

Basilar Arterial Occlusion

- The Basilar Artery supplies the Pons including the Corticospinal Tracts, and the Abducens, Trigeminal, and Facial Nerve Nuclei.
- CONTRALATERAL HEMIPLEGIA can occur if the Corticospinal Tracts are lost.
- MEDIAL or INTERNAL STRABISMUS can occur if the Abducens Nerve Nuclei are lost.
- IPSILATERAL SENSORY LOSS OF THE FACE can occur if the Trigeminal Nerve Nuclei are lost.
- LOSS OF THE MASSETER REFLEX and THE CORNEAL REFLEX can also occur if the Trigeminal Nerve Nuclei are lost.
- BELLS' PALSY and HYPERACUSIS can occur if the Facial Nerve Nuclei are lost.

Commonly Used Neurodiagnostic Tests

LUMBAR PUNCTURE (SPINAL TAP)

- A Lumbar Puncture is used for the diagnostic evaluation of Cerebrospinal Fluid (CSF):
  - Central Nervous System (CNS) Infection (Meningitis, Encephalitis)
  - Neoplastic Invasion of the Subarachnoid Space
  - Multiple Sclerosis
  - Acute Inflammatory Demyelinating Polyneuropathy (Guillain-Barre Syndrome)
  - Neuroimmunologic Disorders
- The usual puncture location is at the level of L2.
- When freely flowing CSF has been obtained, the CSF is measured with a manometer attached to the needle.
- It is then evaluated for cell counts, biochemical and immunologic studies, and microbiologic analysis.

ELECTROENCEPHALOGRAPHY

- Most neurologic imaging studies have supplanted the EEG for localizing anatomic pathology.
- Today, the EEG is used as a pathophysiologic tool able to detect abnormal cerebral function that cannot be visualized radiographically or magnetically.
- The EEG is most commonly used for the evaluation of:
  - Seizures (transient state). While the EEG is useful for the evaluation of seizures, it is not useful for the evaluation of headache or dizziness.
  - Herpes Simplex Encephalitis (evolving condition).
  - Dementia (global disorder).
- Electrodes are placed on specific brain areas and the Encephalographic Waves are analyzed.
- The EEG measures the changing electrical potentials on the scalp in response to controlled stimulation.
- Limitations include:
  - The EEG cannot detect Brainstem Activity well.
  - The EEG is not useful in the diagnosis of Brain Death.

EVOKED POTENTIALS

- Evoked Potentials are used in the evaluation of several sensory modalities including the Visual, Auditory, and Somatosensory Systems.

Visual System

- Visual Evoked Potentials are recorded over the Occipital Region in response to controlled stimulation.
- Can detect Pre- and Post-Chiasmal Lesions.
- Can detect Asymptomatic Optic Neuritis.

Auditory System

- Auditory Evoked Potentials are most commonly used to evaluate Brainstem Structures.
- Five Specific Waves are recorded from the Vestibulocochlear Nerve, the Cochlear Nucleus, the Superior Olivary Complex, the Lateral Lemniscus, and the Inferior Colliculus.
- Auditory Evoked Potentials can detect Physiologic Lesions that are below the limit of resolution of imaging techniques.

Somatosensory System

- Upper Extremity Somatosensory Evoked Potentials are recorded following stimulation of the Median Nerve.
- Lower Extremity Somatosensory Evoked Potentials are recorded following stimulation of the Posterior Tibial Nerve.
- Somatosensory Evoked Potentials can aid in the Diagnosis of Multiple Sclerosis, as Multiple Sclerosis Plaques in the Spinal Cord are difficult to detect using magnetic resonance imaging (MRI).

ELECTROMYOGRAPHY AND NERVE CONDUCTION STUDIES

- These procedures are valuable in the evaluation and diagnosis of Peripheral Nerve and Muscular Disorders.
- They involve the recording of spontaneous, voluntary, and electrically stimulated muscle activity through the use of small intramuscular needle electrodes.
- Nerve Conduction Velocity (NCV) involves electrical stimulation of a Peripheral Nerve. The peripheral nerve's rate of transmission and amplitude of response (to a stimulus) is measured.
- Electromyography, or EMG, and NCV yield data regarding Primary Motor Neuron Disease, Demyelinating vs Axonal Neuropathy, Nerve Root vs Plexus Disorders, and the Localization of a Mononeuropathy.
- Motor Neuron Diseases that can be detected include Amyotrophic Lateral Sclerosis.
- Muscle Disorders that can be detected include Myotonic Dystrophy.
- The most common entrapment syndrome best detected by an EMG is Carpal Tunnel Syndrome.

ANGIOGRAM

- An Angiogram allows the observation of Blood Vessel Size and Shape.
- An Angiogram is a still photographic image.
- It is an invasive procedure that requires the injection of contrast material.
- Anastamoses and Occlusions can be detected using Angiogram technology.

MAGNETIC RESONANCE IMAGING

- MRI uses magnets and radiowaves to detect subtle electromagnetic fields in the brain.
- It is a non-invasive procedure that does not require the injection of contrast material.
- MRI does not expose the patient to ionizing radiation.
- It is valuable for providing images of the heart, large blood vessels, soft tissues, and the brain.
- MRI is more sensitive than Computerized Tomography (CT).
- Useful in the diagnosis of Arteriosclerosis, Arteriovenous Malformations, Vertebral Disc Disease, Spinal Stenosis, Spinal Metastases, Chiari Malformation, Spina Bifida, Spinal Cord Tumors, Demyelinating and Degenerative Brain Disorders, and Brain Hemorrhages, Infarcts, and Space-Occupying Lesions.

MAGNETIC RESONANCE ANGIOGRAPHY

- MRA is a non-invasive method of evaluating vascular structures without the use of contrast material.
- Can detect Aneurysms that cannot be seen by a conventional Catheter Angiography.

COMPUTERIZED TOMOGRAPHY

- CT utilizes the absorption of photons by tissues to generate data that, after computerized post-processing, are presented in a familiar gray scale format.

- Non-invasive.
- While MRIs are more sensitive than CT Scans, CTs are better than MRIs in the detection of Fresh Blood within Cranial Spaces. For this reason, it is the preferred method of imaging a Cerebrovascular Accident (CVA).
- Able to detect Subarachnoid Hemorrhages, Subdural Collections of Blood, Neoplasms, and Vascular Anomalies (Aneurysms and Arteriovenous Malformations).

HELICAL COMPUTERIZED TOMOGRAPHY ANGIOGRAPHY

- Helical CT Angiography uses a three-dimensional technique of data presentation.
- It is able to detect Transient Ischemic Attacks (TIA) in order to prevent a full CVA.

MAGNETIC RESONANCE SPECTROSCOPY

- MRS enables neurologists to make routine direct clinical diagnoses that were previously unavailable by radiologic or clinical tests.
- MRS is used to determine the molecular structure of a compound or to determine the compound's presence.
- It is based on the idea that nuclei of certain atoms have characteristic properties. As a result, signals can be received and displayed as an image.
- Used to differentiate CVAs from Neoplasms.
- Used to make clearer diagnoses when Recurrent Tumor, Neoplasia, and Necrosis cannot be otherwise made.
- MRS is used most commonly as an adjunct to MRI when differential diagnoses cannot be otherwise made.
- MRS has diagnostic value for the evaluation and monitoring of the progression of CVA, Ischemic Injury, Intracranial Tumors, Multiple Sclerosis, Dementias, and Encephalopathies.

POSITRON EMISSION TOMOGRAPHY

- PET produces transverse sectional images of the body by demonstrating the internal distribution of positron-emitting radionuclides.
- Uses harmless radioactive isotopes to reflect the amount to Glucose used by the brain. The more Glucose an area uses, the more active it is.
- PET measures increases in regional cerebral blood flow when people perform specific activities.

SINGLE PHOTON EMISSION COMPUTED TOMOGRAPHY

- SPECT is a scanner that employs radiation detectors to determine the location of a tracer drug in the brain.
- SPECT uses a tracer with a long half-life, making possible studies that involve a prolonged series of scans over a 4-hour period.
- Used to determine which regions of the brain are active during functional activity.

Neurologic Indications for CT

- Acute Cerebrovascular Disease
- Acute Head Trauma
- Extracerebral Tumors (particularly Meningiomas)
- Intracranial Calcification or Osseous Lesions
- Lumbar Spine Presurgical Planning
- Subarachnoid Hemorrhage

Neurologic Indications for MRI

- Cervical Spine Damage
- Demyelinating Disorders
- Head Trauma (after acute phase)
- Inflammatory Disorders
- Intracranial Tumors
- Osseous Vertebral Column Metastases
- Posterior Fossa and Craniovertebral Junction Lesions
- Seizures
- Spinal Cord Lesions

Neurotransmitters: The Neurochemical Basis of Human Behavior

- Neurotransmitters are chemicals that are stored and released from the Terminal Boutons of Neurons. They are released into the Synaptic Cleft.
- Neurotransmitters are either reabsorbed by the Presynaptic Terminal Boutons—called RE-UPTAKE.
- Or Neurotransmitters Diffuse across the Cell Membrane of the Postsynaptic Neuron—called DOWN REGULATION.
- Neurotransmitters can also be destroyed by ANTITRANSMITTERS in the synaptic cleft.
- There are more than 100 Neurotransmitters identified.
- These are Multiple Neurotransmitter Systems in the Brain (and in the Enteric NS).
- Neurotransmitters are EXCITATORY or INHIBITORY depending upon which receptor sites they bind with.
- Neurotransmitters work like the Brain's Brake and Accelerator Systems.

MAJOR NEUROTRANSMITTERS ABOUT WHICH INFORMATION IS KNOWN

- Acetylcholine (ACh)
- GABA (Aminobutyric Acid)
- Glutamate (GLU)
- Dopamine (DA)
- Serotonin (5-HT)
- Norepinephrine (NE)
- Substance P
- Endorphins
- Adenosine

ACETYLCHOLINE (ACH)

- ACh plays an important role in Transmitting Motor Information in the Peripheral NS.
- In the CNS, ACh plays a role in regulating the Autonomic NS.
- In the PNS, ACh facilitates action at the Neuromuscular Junction.
- Excessive ACh at the Motor Endplates of the Neuromuscular Junction can result in Dyskinesia. Dyskinesia is a Hyperkinetic Motor Disorder characterized by Involuntary Muscle Contractions.
- Deficient ACh at the Neuromuscular Junction can result in Paralysis.

Curare Atropine and Australian Fish Toxin

- Curare Atropine is a plant extract that causes paralysis.
- The pharmacologically active ingredient of Curare Atropine is used medically to facilitate skeletal muscle relaxation during anesthesia.
- Curare Atropine is also used by Shaman in certain cultures to anesthetize patients. This causes a zombie-like effect.
- A type of Australian fish releases a toxin that causes complete skeletal muscle paralyses while the individual remains cognitively intact. The toxin blocks the reception of ACh at the Neuromuscular Junction.

Myasthenia Gravis

- Myasthenia Gravis is an Autoimmune Disease that destroys ACh receptors on skeletal muscle membranes. This leads to paralysis.

GABA (AMINOBUTYRIC ACID)

- GABA is a major Inhibitory Neurotransmitter.
- It essentially turns off the function of cells.
- Without Inhibitory Neurotransmitters—like GABA—Brains Cells would fire uncontrollably (as in an Epileptic Seizure).
- There are two known GABA Receptors: GABA-A and GABA-B
- GABA Deficiency is implicated in ANXIETY DISORDERS, INSOMNIA, and EPILEPSY.
- GABA Excess is implicated in MEMORY LOSS and the INABILITY FOR NEW LEARNING.
- Preliminary research suggests that Blocking GABA-B may improve Learning and Memory.

GLUTAMATE

- Glutamate is one of the Major Excitatory Neurotransmitters of the CNS.
- Overactivity of Glutamate may produce Epileptic Seizures.
- There is a delicate balance between Excitation and Inhibition in the brain.
- In Epilepsy, the brain's chemistry has titled too far toward excitation.
- Theorists believe that any disorder—Alzheimer's, Schizophrenia, Parkinson's Epilepsy, CVA—is essentially an imbalance in Neurotransmitter Systems.
- For example, in CVA, theorists believe that it is not just Oxygen Deprivation that kills brain cells, but also a release of Too Much Glutamate.
- When the brain experiences a series of crises—for example, a Head Injury or CVA—Glutamate is released 1000 times more than its normal level.
- This is the organism's way of speeding up a painless death process.
- Physicians, in their attempt to save an individual's life, stop the release of glutamate after it has already killed off an extensive amount of brain cells.
- The release of Glutamate causes the Most Damage in TBIs and CVAs—not the initial injury.
- Researchers are attempting to develop drugs that can immediately prevent the release of Glutamate after a severe brain injury.
- But the drugs would have to be administered in the first hours of the injury, and often patients do not get to the hospital in time.
- Another strategy would be to develop drugs that would cause the release of Glutamate to bind with neighboring neurons without killing them
- Essentially, researchers would be changing the way that Glutamate works in the brain.
- Some researchers believe that in Alzheimer's, the task may be to boost the efficiency of Glutamate to Enhance Memory and New Learning.
- It is believed that Glutamate plays an important role in Learning and Memory, much like GABA.

DOPAMINE (DA)

- The Dopamine System has major effects on the Motor System and on Cognition and Motivation.
- Dopamine does not have an Antitransmitter.
    - It is removed from the Receptor Site by Active Transport or by Diffusion out of the Synaptic Cleft.
    - These processes are slower than Enzymatic Destruction (Antitranmitter Action), and thus the effects of Dopamine last longer than ACh.
- Researchers have identified several sources of Dopamine in the CNS:
    - The Substantia Nigra (in the Midbrain)
    - The Tegmentum (in the Midbrain)
- Loss of Dopamine from the Substantia Nigra is the primary cause of Parkinson's Disease—causing PAUCITY OF MOVEMENT, FESTINATING GAIT, AND MASKED FACE.
- Too Much Dopamine has been implicated in Schizophrenia, causing HALLUCINATIONS, DELUSIONS, DISORGANIZED THINKING, and PAUCITY of THOUGHT.
- Older classifications of Drugs to treat Schizophrenia would often cause Parkinsonian-like Symptoms called TARDIVE DYSKINESIA. Tardive Dyskinesia is a Hyperkinetic Motor Disorder characterized by Involuntary Muscle Contractions: LIP SMACKING, REPETITIVE TONGUE PROTRUSIONS, and BELPHAROSPASMS. Clozapine is a newer classification of Antipsychotic Drug that reduces the affects of Schizophrenia without inducing Parkinsonism.
- Dopamine is also strongly involved in the process of ADDICTION.
- When people take drugs like Heroin, Amphetamines, and Cocaine, it boosts the amount of Dopamine in the brain to unnaturally high levels.
- Dopamine travels through the Brain's Pleasure Center and the Amygdala causing the Sensation of a High.
- This unnatural boost of Dopamine acts like an assault on the brain. To counteract the assault, the brain begins to decrease its natural production of Dopamine.
- This causes Cravings for the drug.

SEROTONIN (5-HT)

- There are 14 identified Serotonin Systems in the CNS.
- Serotonin has been implicated in SLEEP, EMOTIONAL CONTROL and EQUANIMITY, PAIN REGULATION, and CARBOHYDRATE FEEDING BEHAVIORS (or the Binging Behaviors that occur in some Eating Disorders).
- Low levels of Serotonin are associated with DEPRESSION and SUICIDAL BEHAVIOR.
- The Selective Serotonin Re-uptake Inhibitors (SSRI) are a classification of Antidepressants that increase the brain's levels of Serotonin by blocking the Re-uptake of Serotonin in the Presynaptic Neuron.
- Serotonin has also been related to Aggressiveness, Anger, and Violence.
- The Theory holds that the Less Active or Responsive one's Serotonin System is, the More Impulsive and Aggressive one may be.
- Abnormalities in the Serotonin System have been implicated in the White Rage with which some people react to minor irritants.
- The SSRIs reduce some of that Anger, Irritability, and Depression.
- Kramer's Kindling Theory of Depression describes the physiologic process of Chronic, Recurrent Depression.
- The first three or four Depressive Episodes (which may be minor) cause a permanent change in the Serotonin System.
- Like a Domino Effect, with each successive Depressive Episode the synaptic vessels produce and release less and less Serotonin.
- The Serotonin System may never return to normal without intervention.

The Use of SSRIs in the Treatment of Various Disorders

- The SSRIs have been effective in the treatment of disorders other than depression and anger.

Obsessive Compulsive Disorder (OCD)
- OCD involves obsessions or continuous thoughts and compulsive behaviors that attempt to alleviate the obsessive thoughts.
- OCD involves survival behaviors taken to an extreme:
  - Cleanliness: (eg, hand washing and washing one's clothing excessively).
  - Touching behaviors: Touching Behaviors usually involve superstitious thinking. "I have to touch the doorknob 5 times before I can go to sleep."
  - Checking behaviors: (eg, checking the alarm to see if it is set 10 times before going to sleep; checking the door to make sure it is locked repeatedly).
  - Hoarding; (eg, saving everything in case one may need it one day; piling up groceries or clothing to make sure one always has enough).

Eating Disorders
- In Bulimia, the SSRIs decrease the binging behavior.
- In Anorexia, the SSRIs decrease the fasting of food or food restriction.
- The SSRIs have been more effective in the treatment of bulimia than anorexia.

Kleptomania (A form of OCD—Hoarding)
- The SSRIs have been effective in decreasing Kleptomania.

Panic Disorders
- Panic Disorder involves a sudden sensation of doom or non-reality based fear, a sense of lost personal identity and the perception that familiar things are strange or unreal.
- These symptoms are accompanied by increased heart rate and blood pressure—increased sympathetic NS functions.
- Panic Disorder has been traced to an oversensitivity in one of the serotonin systems.
- Paxil is an SSRI that effectively decreases panic disorder and social phobia because it increases serotonin levels.
- But patients must take higher levels of Paxil (40-60 mg per day) to experience a decrease in panic disorder, anxiety and social phobia.
- Paxil also reduces symptoms rather than completely alleviating them.

NOREPINEPHRINE (NE)

- NE plays a role in Active Surveillance of one's surroundings by increasing attendance to sensory information from the environment.
- NE is essential in producing the Flight/Fight Reaction to stress.
- NE does not have an Antitransmitter.
- It is removed from the receptor site by Active Transport or by Diffusion out of the Synaptic Cleft.
- These processes are slower than Enzymatic Destruction (or Antitransmitter Activity), and thus the effects of Norepinephrine lasts longer than ACh.
- Overactivity of the NE System produces FEAR and PANIC (through the action of Cortical and Limbic Regions).
- Beta Blockers—such as Propanolol—prevent NE from binding to Beta Receptors. This prevents Sweating, Rapid Heart Beat (Tachycardia), and other Sympathetic NS signs that may occur in stressful situations.
- Musicians, Actors, and Public Speakers have been known to take Beta Blockers before performances to reduce Sympathetic NS signs and to enhance calmness.

SUBSTANCE P

- Substance P acts as a Neurotransmitter in the Nociceptive Pathway. It is classified as a Peptide.
- The Nociceptive Pathway mediates the sensation of pain. It is found in the Dorsal Horn of the SC, Substantia Nigra, Amygdala, Hypothalamus, and the Cerebral Cortex.

ENDORPHINS

- Endorphins work conjointly with Substance P to act as Pain Modulators.
- Endorphins have been described as the Body's Natural Morphine.
- The primary action of Endorphin is the Inhibition of Nociceptive Information.

ADENOSINE

- Adenosine is an Inhibitory Neurotransmitter.
- The release of Adenosine during sleep facilitates Sleep Maintenance.
- Adenosine also causes the Arousal Threshold to be Increased—this means that it takes more to cause arousal or to wake the individual from sleep.
- Adenosine also facilitates a return to sleep if the individual is awakened.

NEUROTRANSMITTERS INVOLVED IN WAKEFULNESS/CONSCIOUSNESS

- Neurotransmitters involved in Wakefulness and Consciousness include Norepinephrine, Dopamine, Acetylcholine, Histamine, and Glutamate.
- Neuropeptides involved in Wakefulness and Consciousness include Corticotropin Releasing Hormone (CRH), Thyrotropin Releasing Hormone (TRH), and Vasoactive Intestinal Peptide (VIP).
- Deficiency in one or more of these substances can produce Somnolence (sleep).
- Growth Hormone is also deficient in Chronic Fatigue Syndrome (CFS) and Fibromyalgia. Theorists believe that one cause of CFS and Fibromyalgia is a lack of Restorative Sleep—Deep Stage 4 Sleep.

NEUROTRANSMITTERS CAN OVERLAP IN FUNCTION

- Several different Neurotransmitters can regulate the same behaviors.
- This redundancy in the regulation of same functions is seen throughout the CNS and the PNS.
- It is likely the Nervous System's way of creating a fool-proof system—akin to having back-up systems if something goes wrong. This likely developed as an evolutionary safe-guard for the survival of human beings.
- DOPAMINE and SEROTONIN appear to regulate functions and disorders regarding the experiences of ANGER, DEPRESSION, ANXIETY, and ADDICTIVE BEHAVIORS.
- NOREPINEPHRINE also appears to regulate ANGER and ANXIETY.
- GABA and GLUTAMATE have shared roles in MEMORY and LEARNING.
- ACETYLCHOLINE and DOPAMINE have shared roles in the regulation of MOTOR ACTIVITY.

The Neurologic Substrates of Addiction

THE MESOLIMBIC DOPAMINE SYSTEM

- During Cravings for an Addictive Substance PET Scans reveal activation in:
  - The Amygdala (in the Temporal Lobe; part of the Limbic System)
  - The Anterior Cingulate Gyrus (located right above the Corpus Callosum)
  - The Frontal Aspects of the Temporal Lobes
- These structures form the Mesolimbic Dopamine System. This system is responsible for the Sense of Pleasure—referred to as the Brain's Pleasure Center.
- When the Amygdala is Lesioned, individuals no longer link Pleasure to its causes.
- Repeatedly dosing the brain with Addictive Drugs is akin to a Chemical Assault that alters the Structure of Neurons in the Pleasure Center.
- Cocaine floods the Neurons with abnormal levels of Dopamine. The cells attempt to adapt to this repeated assault by becoming less responsive to Dopamine.
- These changes starve Brain Cells of Dopamine, triggering a Craving for the Addictive Substance.
- PET Scans of Cocaine Rehabilitation Patients taken over several weeks after they have stopped using the drug show a drop in the Pleasure Center's ability to receive Dopamine.
- This is why Addicts need greater amounts of the drug to attain the same quality of high they experienced on the first occasions they took the drug.
- This reduction in the Pleasure Center's ability to receive Dopamine will last 1 to 1½ years after withdrawal.
- The highest risk of relapse for Cocaine Addicts occurs during the 3rd or 4th week after they have stopped taking the drug. This is when the lowest levels of Dopamine occur in the Mesolimbic Dopamine System.
- The Brains of Addicts almost always return to normal after a year without the drug—though not completely. Dopamine Receptor Cells may never return to normal.
- If an individual can remain abstinent for 1 year, he or she has weathered the period of greatest vulnerability of recidivism.

ADDICTIONS TO ACTIVITIES

- Addiction is any repeated behavior that produces a chemical change in the Brain's Pleasure Circuit, thus compelling the individual to continue the Addicting Behaviors.
- In addition to Substance Use, Addictive Behaviors include:
  - Eating Disorders (Anorexia, Bulimia, Compulsive Binge Eating)
  - Work (Workaholism)
  - Relationships/Sex
  - Gambling

- Shopping
- Kleptomania
- These repeated behaviors cause a Chemical Change in the Brain that produces a Sense of Well-Being.

Addictions Are Cross-Addicting

- Individuals often recover from one addiction only to go to another.
- For example, Alcoholics often recover from their substance abuse to replace it with 12-Step Programs. The Meetings become the Addiction.

Addictive Behaviors Are Likely Genetically Based

- There is evidence that Addictive behaviors are Genetically Based.
- The lack of Serotonin/Dopamine that causes Addictive Behaviors tends to run in families.
- The type of Addiction that an individual gravitates toward is often culturally based and influenced by the family of origin.
- For example, it is common to observe that children of Alcoholics often have OCD or Eating Disorders, and vice versa.
- As a result of Cultural Influences, Eating Disorders, or Body Dysmorphic Disorders express themselves differently in males and females.
  - While Females may engage in Anorexic or Bulimic Behaviors to stay slim (because it is culturally valuable to be a thin female), Males may overly engage in Body Building (because the culture values men who appear large and strong).

Addictions Are Obsessions

- Obsessions are recurrent ideas that invade consciousness.
- Obsessions are mediated by Brain Loops that may originate in the Basal Ganglia.
- The abnormality in the Brain Loop involves recurrent electrical brain waves that will not stop firing—the Brain Loop will not shut off.
- There is a link between Obsessive Thoughts and having to act out the Compulsive Behaviors—in which the Basal Ganglia may play an important role.
- The Basal Ganglia are a part of the Primitive Survival Centers in the Brain.
- Obsessions involve ideas and behaviors that are linked with Survival and Control: Food, Money, Cleanliness.

Obsessive-Compulsive Disorder Is a Component of All Addictions

- OCD is believed to be caused by the same lack of Serotonin/Dopamine that is believed to cause Addictions.
- OCD often involves Obsessions with:
  - The Symmetry of Objects
  - Checking Behaviors
  - Cleanliness
  - Counting
  - Invasive Thoughts about Sex, Religion, Catastrophic Events
  - Invasive Music that Runs Through the Mind Repetitively
  - Co-Dependency is an Obsession

Serotonin as an Obsession Blocker

- Serotonin helps to shut off the Brain Loop that causes the Addictive Behaviors.
- The more Serotonin the Brain has available to it, the less Obsessive Behaviors occur.
- Some Theorists label Serotonin the Peace of Mind Chemical.
- The SSRIs build up Serotonin in the Brain and are often referred to as Obsession Blockers.

Learning and Memory

EARLY THEORIES ABOUT MEMORY

- Two decades ago, theories of Memory were based on a Computer Model.
- Memories were compared to Computer Files that could be placed in Storage and pulled up into consciousness when needed.
- Today, more is known about how Memory works. Information has been gained through experimental research and research on patients with cerebral damage.

MEMORY IS NOT A FACTUAL-BASED RECORD OF REALITY

- Our Memories are not passive or literal recordings of reality.
- Example: Specific events are remembered differently by various members of one family.
- Some members have forgotten the memory while others adamantly recall their occurrence.
- People rarely recall all of the details in an event accurately.
- Individuals often recall occurrences that made general sense or fit their expectations of what should have happened, but were not actually part of the original event.
- False Memories have become highly debated in recent years as reports of early childhood abuse are questioned. Some have asked if False Memories are the product of psychological techniques that induce memories of events that never occurred.

MEMORIES ARE ENCODED BY BRAIN NETWORKS

- An individual's experience is encoded by Brain Networks involving Multiple Cerebral Structures. There is no single anatomical structure that alone deals with memory functions.
- PET Scans have enabled researchers to observe the brain in action while people are remembering specific events.
- Memories are encoded by Brain Networks whose connections have already been shaped by previous encounters with the world.
- This pre-existing structure powerfully influences how individuals encode and store new memories. It influences what an individual will remember and why.
- Memories that have a strong emotional significance appear to be more readily remembered. Perhaps because the Hippocampus and the Amygdala (both Limbic System Emotional Centers) play an important role in the storage of Long-Term Memories (LTM) that have a strong emotional component.
- But these LTMs having an emotional significance are also subject to distortion based on an individual's psychological interpretation of an event.
- Human Memory is thus predisposed to corruption by suggestive influences. Memories more likely involve distortions of reality rather than being factual snapshots or a video.

Types of Memory

Semantic Memory
- Semantic Memory involves the recall of Facts—dates of people's birthdays, state capitols, the names of presidents.
- Semantic memory includes the definitions of words and how to use the rules of grammar.

Episodic Memory
- Episodic Memory involves Significant Events that happened to an individual—the first day of school, one's wedding, the birth of a child.

Procedural Memory
- Procedural Memory involves the recall of steps involved in specific tasks. For example, knowing the steps involved in fabricating a wrist cock-up splint. Or knowing the steps involved in changing a flat tire.

Explicit Memory
- Explicit Memory involves knowledge that an individual consciously knows she has acquired. For example, an anatomist knows that he has acquired knowledge of the human body's musculoskeletal system.

Implicit Memory
- Implicit Memory involves knowledge that the individual does not consciously know he has acquired. For example, someone with amnesia may not consciously know she has acquired knowledge of how to play the piano. But when seated at the piano, she can play fluently.

Source Memory
- Source Memory involves the recall of how information is learned. For example, an anatomist may recall that he learned specific information about the human shoulder joint from a specific dissection class.
- Often, people remember facts but cannot recall how they came to learn those facts. An example involves the rehearsal of legal witnesses by lawyers. If lawyers over-rehearse a witness, the witness may confuse rehearsed information with the events that they actually observed.

Field Memories vs Observer Memories
- Field and Observer Memories refer to an individual's own perspective of herself within the remembered event.
- In a Field Memory, individuals do not see themselves in the memory. The memory is remembered as though the individual is seeing it through his or her own eyes.
- In an Observer Memory, the individual has incorporated his own image into the remembered event. The individual has altered the original scene—because the original perspective of an event is always viewed from a field perspective. Individuals are not outside of their bodies, viewing an event and observing themselves in the event.
- More Recent Memories are remembered from a Field Perspective, while Older Memories are remembered from an Observer Perspective.
- This is a phenomenon of Time. Individuals tend to incorporate themselves into the remembered event as time goes by.

Retrograde Amnesia vs Anterograde Amnesia
- Retrograde Amnesia is a common consequence of Brain Damage.
- Involves the Loss of one's Entire Personal Past. This loss occurs after the injury.
- Past Memory is often recovered after injury at some point.
- LTM often returns to patients as they recover from Brain Injury.
- Anterograde Amnesia also occurs after an injury. It involves (a) a decreased ability to remember ongoing day-to-day experiences and (b) a decreased ability to Encode Short-Term Memories (STM) into Long-Term Storage. LTM remains intact.

The Encoding Process
- The Encoding Process involves the actual steps the Brain uses to turn an event into a Stored Memory that can be recalled.
- STM only lasts for seconds unless it is encoded and stored.

- Another name for STM is Working Memory.
  - Working Memory is a function of the Prefrontal Cortex.
  - It consists of Moment-to-Moment Awareness.
  - Working Memory also plays a role in the Search and Retrieval of Archived Information.
- LTM is a function of the Hippocampus and other Cortical Areas that researchers are only beginning to understand.
- Example: STORAGE OF THE NAMES OF TOOLS is a function of the Medial Temporal Lobe and the Left Premotor Area.
  - Humans name Tools through the Association of the Sounds that they make. This is a function of Audition and the Temporal Lobe.
  - Humans also name Tools by their Actions. This is a function of the Premotor Area.
- The STORAGE FOR THE NAMES OF ANIMALS is primarily a function of the Medial Occipital Lobe.
  - Humans categorize Animals primarily by their Appearance (a function of the Occipital Lobe).
  - Humans also categorize Animals by the Sounds they make (a function of the Temporal Lobe).
  - The way the Animal Feels to one's Touch is a function of the Parietal Lobe.
  - Storage for the Movement Characteristics of Animals appears to be a function of the Premotor Area.
- When patients have difficulty naming Actions (eg, swimming, driving), it usually indicates Prefrontal Lobe Damage. The Prefrontal Lobe is where words of Action are Processed and where Motor Plans are Stored.
- Memories of Single Objects are Fragmented into Pieces that are Stored in Different Regions of the Brain.

The Frontal Lobes consist of Distinct Sub-regions that play important roles in Elaborative Encoding, Strategic Retrieval, Working Memory, and Recall of Source Information.

Specific Regions in the Parietal, Occipital, and Temporal Lobes participate in the Storage of Different Attributes of LTMs.

The Cortical Areas cooperate closely with Structures in the Hippocampus to allow individuals to remember ongoing day-to-day experiences.

ESTABLISHING A DURABLE MEMORY

- Incoming information must be Encoded thoroughly, or deeply, by Associating it with Meaningful Knowledge that Already Exists in Memory Storage.
- The Human Memory System is built so that individuals are more likely to remember what is most important to them.
- The Amygdala and Hippocampus play an important role in Encoding Memories that have Emotional Significance.

ENCODING AND MNEMONIC DEVICES

- Mnemonic Devices are Techniques that assist Learning and Recall of Learned Material.

Method of Loci

- The Method of Loci is a Mnemonic Device involving the creation of a Visual Map of one's house.
- For example, if one wanted to remember to buy soda, potato chips, and soap, one would use the rooms of one's house and visualize the kitchen with spilt soda on the floor, the bedroom with potato chips scattered on the bed sheets, and the bathroom tub filled with soap bubbles.
- At the store, one would then take a mental walk around one's house and see what is in each room.
- The Cognitive Act of creating a Visual Image and Linking it to a Mental Location is a form of DEEP ELABORATIVE ENCODING.

Acronyms

- ACRONYMS are another form of Mnemonic Device.
- There are acronyms that students use to remember the Cranial Nerves and the Bones of the Skeleton.

DAMAGE TO FRONTAL LOBE REGIONS AND ENCODING PROBLEMS

- Patients who have sustained damage to the Frontal Lobe Regions often have Encoding Problems.
- They fail to organize and categorize new information as it comes into memory.
- The Left Inferior Prefrontal Cortex plays an important role in Deep or Elaborative Encoding.

ENCODING OF NOVEL EVENTS AND THE HIPPOCAMPUS

- The Hippocampal Region becomes activated when individuals Encode Novel Events.
- Individuals may activate the Hippocampus in response to novel stimuli to see if they have already developed and stored associations to that stimuli.
- When associations are found (or if none are found), the Prefrontal Cortex becomes activated in preparation to Encode the Novel Stimuli.
- Research with brain injured patients suggests that damage to the Hippocampus can produce Severe STM Loss. This is because the Hippocampus can no longer Encode New Memories. However, LTM usually remains intact.

ENGRAMS

- Engrams refer to the Memory Trace, or Neural Pathway that has been created for the recall of each stored memory.
- Engrams are Theoretical Constructs.
- More likely, there is No Single Pathway that, when activated, allows individuals to recall a specific memory.

RECONSTRUCTION OF MEMORY

- The Brain engages in an act of Construction during the Retrieval Process.
- There is no single location or area of the brain that contains an Engram of a particular past experience.
- Instead, the Perceptual Regions of the Brain that are concerned with Sight and Sound merge with the Convergence Zones, or the Posterior Multimodal Association Area.
- The Posterior Multimodal Association Area contains codes that bind sensory fragments to one another and to pre-existing knowledge.
- The Retrieved Memory is a Temporary Construction of information from several distinct brain regions—a Construction having many contributors.
- Neisser's Analogy: Retrieving a Memory is akin to Reconstructing a Dinosaur from Fragments of Bone.
- Researchers believe that the Medial Temporal Lobe (a region within the Posterior Multimodal Association Area) contains a kind of Index that points to the location of different kinds of information that are stored in separate cortical areas.
- The Index is needed to keep track of all sights, sounds, and thoughts that together comprise an event until the memory can be held together by direct connections between the cortical regions themselves. The Index is then no longer needed to recall a specific event.
- When humans remember events, it is like a giant jigsaw puzzle assembled by many cerebral contributors.

THE CONSOLIDATION PROCESS

- At the psychological level, LTM Consolidation occurs in part because people talk and think about their past experiences.
- The older the memory, the greater the opportunity for such past event rehearsal.
- Thinking and talking about a past experience promotes the direct connections between cortical storage areas.
- Once an experience has been repeatedly retrieved, it becomes consolidated and no longer depends upon the integrity of the Medial Temporal Lobe Structures to act as an Index.
- Sleep also plays a role in the Consolidation Process—particularly Rapid Eye Movement (REM) Sleep.
- During Sleep, the Hippocampus plays back a recent experience to areas of the cortex where it will eventually be stored.

THE PASSAGE OF TIME AND DECREASING MEMORY

- As time passes, humans Encode and Store new experiences that interfere with their ability to recall previous ones.
- For about 1 year after the occurrence of an event, it can be accessed readily—with virtually any cue.
- As more time elapses and the Memory becomes blurry, the range of cues that elicit a specific remembered event progressively narrows.
- This means that when an individual suddenly and unexpectedly recovers a seemingly forgotten memory, it may be because he or she has stumbled upon a retrieval cue that matches perfectly with the faded or blurred memory.

MEMORY LOSS IN PATIENTS WITH HEAD INJURY

- Most patients who have sustained a Head Injury in an accident cannot remember the accident itself.
- This likely occurs because the accident was never encoded from STM to LTM. The individual became unconscious before the Encoding Process could take place.
- LTM tends to remain intact (after the period of Retrograde Amnesia resolves).
- This is because Old Memories are subject to a Long-Term Consolidation Process that allows them to become more resistant to disruption over time.
- People with permanent memory deficits resulting from damage to the Hippocampus are better at remembering the Distant Past than the Recent Past.
- The Hippocampus plays an important role in the Consolidation of LTM.
- Patients with Parietal Lobe Damage commonly forget once familiar spatial layouts and have difficulty navigating routes that they used to travel with ease.

The Neurologic Substrates of Emotion

- Research regarding how the Brain processes Emotion has been advanced in recent years by PET Scans. PET Scans have enabled researchers to identify which brain regions become activated when people experience specific emotions.
- Much like Memory, researchers used to believe that Emotion was processed by one single cerebral system—largely the Limbic System. Today, researchers contend that Emotions are processed by Multiple Networks of Cerebral Structures.

THE ROLE OF THE LEFT VS. THE RIGHT HEMISPHERE IN EMOTION

- The Left and Right Hemispheres have different roles in the processing of Emotion.
- Both Hemispheres act in a Mutual Cross-Modulatory Relationship—a relationship much like Reciprocal Inhibition in which each hemisphere keeps the other's innate emotional response in check.
- The two Hemispheres are Specialized for the processing of different kinds of Emotions—just as they are specialized for different kinds of cognitive, sensory, perceptual, and motor skills.

THE MAJOR NEUROANATOMICAL STRUCTURES INVOLVED IN EMOTION

- The primary neuroanatomical structures involved in the processing of Emotion include the Prefrontal Cortex (in both the Left and Right Hemisphere) and the Limbic System (including the Amygdala, Hippocampus, and the Cingulate Gyrus).
- Secondary Structures in the processing of Emotion include the Thalamus, the Anterior Insular Region (of the Temporal Lobe), and the Septum Pellucidum.

THE PREFRONTAL AREAS OF THE LEFT VS. THE RIGHT HEMISPHERE

- PET Scans have revealed that the activation of the Left Prefrontal Area coincides with Positive Emotions and a Sense of Well-Being.
- Activation of the Right Prefrontal Area coincides with Emotions of Agitation, Nervousness, Distress, Anxiety, Sadness, and Depression.
- Stimulation of the Left Prefrontal Area through Transcranial Magnetic Stimulation (TMS) appears to enhance one's mood.
- Stimulation of the Right Prefrontal Area by TMS produces depressed moods in people.
- The effect of TMS on an individual's mood only lasts for several hours.

THE RIGHT AND LEFT HEMISPHERES MAY DIVIDE NEGATIVE AND POSITIVE EMOTIONS

- When people report feelings of Sadness, Depression, and Anxiety, their Right Prefrontal Lobe shows increased electrical activity on SPECT Scans.

- Emotional States of Happiness and Enjoyment increase the electrical activity in the Left Prefrontal Lobe.
- Infants who are prone to distress when separated from their mothers show increased activity in the Right Prefrontal Lobe. People who identify themselves as Pessimists also show increased activity in the Right Prefrontal Lobe.
- People who were at one time Depressed (but who do not report feelings of Depression in the present) show decreased Left Prefrontal Lobe Activity when compared to people who never experienced Depression. This finding reinforces Kramer's Kindling Theory of Depression—that each Depressive Episode produces permanent changes in the Brain's Structure and Chemical Balance.

TWO SYNDROMES RELATED TO THE SITE OF BRAIN INJURY

Left Prefrontal Lobe Damage
- Patients with Left Prefrontal Lobe Damage tend to be Emotionally Labile, Depressed, and Despondent. This is sometimes referred to as a Catastrophic Reaction.
- Brain Damage to the Left Hemisphere leaves the Intact Right Hemisphere to become the Dominant Hemisphere.

Right Prefrontal Lobe Damage
- Right Prefrontal Lobe Damage often leaves patients with an indifference to their impairment (Anosagnosia) and sometimes a reported state of Euphoria or Well-Being.
- This occurs because Brain Damage to the Right Hemisphere leaves the Left Hemisphere to become Dominant.
- The Right Hemisphere may have an Innate Emotional Bias toward experiences of Negative Emotion and toward a world view of Pessimism.
- Brain Damage to the Left Hemisphere releases the Right Hemisphere's Anxious-Depressed Emotional Perspective. This may account for the clinically observed Catastrophic Reaction.
- Conversely, Damage to the Right Hemisphere sets free the Optimist that lives within the Left Hemisphere. This may account for the optimistic denial of disability and a disowning of impaired appendages.

THE LEFT HEMISPHERE MODULATES THE EMOTIONAL RESPONSE OF BOTH HEMISPHERES

- It appears that the Left Hemisphere may modulate the emotional responses of both hemispheres.
- Patients with Bilateral Hemispheral Damage tend to become Depressed—as in the case of Unilateral Left Hemisphere Damage. It is as though the co-existing Right Hemispheral Injury had not occurred.
- When both the Left and Right Hemispheres are Bilaterally Damaged, the Left Hemisphere cannot alleviate the Right Brain's Negative Emotional Responses and the patient tends to become depressed.
- The Left Hemisphere may play a role in balancing an individual's Emotional Equanimity—so that, under normal circumstances, the Right Brain's propensity toward Depression does not take over.
- In patients with Chronic Episodic Depression, it appears that the Left Brain's ability to mitigate the Right Brain's Emotional Negativity becomes decreased with each Depressive Episode.

THERAPEUTIC SIGNIFICANCE

- If Depressed Patients exhibit heightened activity in the Right Hemisphere, one way to enhance mood may be to provide activity that Stimulates the Left Brain's Function.
- This may account for the high degree of effectiveness of Cognitive Therapy with Depressed Patients. Research has shown that Cognitive Therapy—concomitantly administered with drug therapy—has been found to be more effective than drug therapy alone.
- Cognitive Therapy challenges the patient to assess his or her cognitive mindsets in order to determine if those thought patterns are reality based. If one's thought patterns are found to be non-reality based, the therapist challenges the patient to make needed mental set changes. These activities require Left Brain Cognitive Functions.
- Occupational Therapy that provides depressed patients with activity requiring Left Brain Functions may also be an effective treatment for Depression (if administered conjointly with drug therapy).
- It is also likely that Workaholism has the same effect as stimulating Left Brain Functions. People who are workaholics may be using work in an attempt to self-medicate an underlying Depression.

ORBITOFRONTAL VS. DORSOLATERAL FRONTAL LOBE LESIONS

- There are two further Emotional Syndromes related to the site of Brain Injury.

Orbitofrontal Lobe Lesions

- Patients with Orbitofrontal Lobe Lesions exhibit Impulsiveness and Disinhibition.
- The Orbitofrontal Region mediates the Brain's Executive Functions. When the Orbitofrontal Region is Lesioned, the Executive Functions become impaired.
- Patients exhibit Poor Judgment, Risk-Taking Behaviors, Rowdiness, and Socially Inappropriate Behaviors.

Dorsolateral Frontal Lobe Lesions

- Patients with Dorsolateral Frontal Lobe Lesions exhibit Decreased Drive and Motivation, Lethargy, and a Flat Affect.
- They rarely initiate activity or display emotion. It is as though their Emotional Center has become neutral to the point of inactivity.

THE AMYGDALA'S ROLE IN EMOTION

- The Amygdala has several roles in the Processing of Emotion.
- The Amygdala is part of the Mesolimbic Dopamine System, or the Pleasure Center of the Brain.
- Patients with Lesions to the Amygdala do not experience pleasure from people and activities that once provided them with an internal sense of happiness.
- Patients who have Bilateral Lesions to the Amygdala no longer experience the Emotion of Fear.
- Such patients have difficulty recognizing dangerous situations and often engage in risk-taking behavior as a result.
- The Amygdala works cooperatively with the Anterior Cingulate Gyrus, Prefrontal Lobe Regions, and the Right Hemisphere to assess whether situations are dangerous.
- This circuit generates a response that is transmitted along two pathways for feedback about the internal and external environment.
    - One Pathway travels to the Sympathetic Nervous System to check the body's sympathetic response to a stimulus. Messages from the Sympathetic Nervous System are then sent to the Secondary Somatosensory Area (SS2).
    - SS2 then sends this information to the Thalamus and the Prefrontal Lobe for cognitive decisions regarding the appropriate response to the stimulus in question.
    - Example: An individual hears the roar of a lion.
    - Her Sympathetic Nervous System becomes activated.
    - The Sympathetic Nervous System information travels to SS2.
    - SS2 then sends this information back through the Thalamus and to the Prefrontal Cortex.
    - The Prefrontal Cortex, observing that the lion is in a confined zoo area, overrides the Amygdala's initial fear response.

PATIENTS WITH LESIONED AMYGDALAS CANNOT RECOGNIZE FEAR IN OTHERS

- When presented with a chart of faces with different emotions, patients with lesioned Amygdalas are unable to recognize the emotion of fear on the observed faces. However, they are able to identify the expressions of happiness, sadness, and anger.
- They are also unable to produce the facial expression of fear when asked to do so.

THE AMYGDALA MAY PLAY A ROLE IN THE RECOGNITION OF SOCIAL AND EMOTIONAL CUES

- There is evidence that individuals with Autism have impairment within the Limbic Structures—particularly the Amygdala.
- The inability to interpret social cues and to generate an appropriate emotional response to those cues (eg, empathy, sadness, or happiness for others) is significantly impaired in people with Autism.

- The Anterior Insular Region, the Amygdala, and the Hippocampus appear to be involved in attaching Emotional Significance to Thoughts and Memories. If any of these brain regions are lesioned, the individual has difficulty identifying important and meaningful experiences.
- Such individuals also have difficulty recognizing meaningful events for others and tend to react without emotion in response to another's discussion of emotionally significant experiences. This, too, is a characteristic of Autism.

The Amygdala and Post-Traumatic Stress Disorder

- In PET Scans of individuals who experience PTSD, or Post-Traumatic Stress Disorder, the Amygdala becomes highly activated—perhaps overactive.
- Individuals with PTSD report feeling that the traumatic event is vividly occurring all over again—as though the Amygdala and other Limbic System Structures were flooding the Visual, Auditory, and Somatosensory Cortices with the painful memories.
- Another phenomenon of PTSD is that the Language Areas of the Brain shut down during the PTSD experience. Individuals experiencing PTSD episodes report that they cannot put their emotions regarding the event into words (Alexithymia).
- Without the ability to put language to their emotions, they cannot integrate the Right Hemisphere's Emotional Response to the event with the Left Hemisphere's ability to cognitively analyze the event.
- The traumatic event continues to be experienced as an overwhelming somatic experience that has a life of its own and that invades one's consciousness without modulation from the Left Hemisphere.
- The treatment for PTSD has not been significantly effective. This may be related to the finding that emotional memories involving extreme fear are permanently encoded into the Limbic Structures. While such memories can be suppressed, they can never be fully erased.
- The goal of treatment is to mitigate the flooding experiences and to help the individual to use cognitive and emotional strategies to subdue the PTSD when it occurs.

The Limbic System's Role in Anxiety, Panic Attacks, Phobias, and Obsessive Compulsive Disorder

- PET Scans of individuals with Anxiety, Panic Attacks, Phobias, and OCD reveal heightened activity in the Limbic System Structures—particularly the Anterior Cingulate Gyrus, the Amygdala, and the Anterior Temporal Cortex (the location of the Parahippocampal Gyrus). Other structures also include the Insula and the Posterior Orbitofrontal Cortex.
- This Circuit is referred to as the Worry Circuit.
- The Locus Coeruleus (part of the Brainstem that rouses the body to action by secreting Norepinephrine) also becomes activated during Anxiety, Panic Attacks, Phobias, and OCD.

The Septal Area and Anger

- Another brain region that appears to mediate Anger is the Septal Area—a region on either cerebral hemisphere comprising the subcallosal area and the Septum Pellucidum. THe area has Olfactory, Hypothalmic, and Hippocampal Connections.
- The role of the Septal Area is to provide a means of communication between the Limbic Structures and the Diencephalon (particularly the Thalamus and the Hypothalamus).
- If the Septal Area is stimulated with an electrode in a cat, the cat lashes out with rage.
- PET Scans reveal heightened activity in the Septal Area when individuals experience Anger.
- A lesion to the Septal Area in animals produces rage and hyperemotionality.
- In Humans, destruction of the Septal Area gives rise to Emotional Overreactions to stimuli in the environment.

Differences in the Neurologic Substrates of Emotion Between Males and Females

- The Brains of Males and Females generate specific Emotions using different Patterns of Brain Activity.

- When females feel sadness, they experience greater activity in the Anterior Limbic System than do males.

- According to some research studies, females also tend to experience a more profound sadness than do males. This may reinforce the research finding that females experience twice the risk of Depression than men.

- Because females activate both hemispheres during the experience of significant emotions (as observed in PET Scans), they may be more able to put words to their emotions—a process that requires using the Left Hemisphere Language Areas to associate words with one's emotions. As a result, they may be more aware of their emotions than males.

- Males do not as readily integrate both hemispheres during the experience of significant emotions. This may predispose males to greater difficulty attaching words to their emotions—or even being aware of their emotions.

The Aging Brain

EPIDEMIOLOGY

- In developed nations, the leading cause of Senile Dementia (loss of Memory and Reason in the Elderly) is Alzheimer's Disease.
- Other Diseases of High Incidence in the Elderly include Parkinson's Disease and Multiple Cerebrovascular Accidents (CVAs).

PET SCANS OF AGED BRAINS

- PET Scans of Aged Brains reveal:
  - Cortical Atrophy
  - Broadening of the Sulci
  - A Decrease in the Size of the Gyri
  - A Widening of Ventricular Cavities

AGE-RELATED DAMAGE

- Age-Related Damage depends upon:
  - The Age of Onset of the Damage
  - The Type and Extent of Physical Brain Alterations
  - One's Medical History
  - One's Genetic History
  - One's Use of Brain Structures throughout the Lifespan
- Most age-related structural and chemical changes become apparent in Late Middle Life—the 50s and 60s.
- Some age-related changes become pronounced after age 70.

AGE-RELATED DAMAGE IN THE DIENCEPHALON AND BRAINSTEM

- Certain subsets of cells and areas of the brain are more prone to age-related damage than others.
- As people age, their overall number of Brain Neurons Decreases.
- Few neurons disappear from the Hypothalamus—a structure that regulates the secretion of certain Hormones by the Pituitary Gland.
- Conversely, many neurons disappear from the Substantia Nigra and Locus Coeruleus (both located in the Brainstem). Parkinson's Disease can destroy 70% or more of the Neurons in these areas.
- In the healthy elderly, 30% to 40% of neurons disappear from the Substantia Nigra and Locus Ceruleus.

AGE-RELATED DAMAGE IN THE LIMBIC SYSTEM

- The Hippocampus can lose 5% of Neurons each decade in the second half of life. Twenty percent of Hippocampal Neurons can be lost as a result of the Normal Aging Process.
- Large Neurons tend to be more affected than Short Neurons. Large Neurons have been found to shrink in the Hippocampus and Cerebral Cortex.
- Cell Bodies and Axons also degenerate in certain Acetylcholine-Secreting Neurons. These types of Neurons project from the Forebrain to the Hippocampus.
- Some Neuronal changes may represent attempts by surviving Neurons to Compensate for Loss or Shrinkage of other Neurons and their projections.
  - Studies have found an increase in the Net Growth of Dendrites in the Hippocampus and some areas of the Cortex.
  - This occurs in the 40s and 50s.
  - In late life (80s and 90s), the Net Growth of Dendrites in these areas decreases.
  - This may suggest that the Brain is capable of Dynamic Remodeling of its Neuronal Connections.
  - This may also suggest that Therapy can augment this type of age-related Neuronal Plasticity.

NEURONS AND CELL BODIES UNDERGO AN ALTERATION OF THEIR INTERNAL ARCHITECTURE

- The Cytoplasm of cells in the Hippocampus and the Cortex (both critical for Learning and Memory) can begin to fill with Protein Filaments known as Neurofibrillary Tangles.
- An Abundance of these Tangles is believed to contribute to Alzheimer's Disease.
- Small accumulations of Neurofibrillary Tangles in the Healthy Elderly are not well understood.
- Neuronal Cytoplasm in many parts of the brain become increasingly speckled with many granules containing Lipofuscin—a fluorescent pigment.
- Researchers do not understand whether these granules harm cells or are simply concomitant parts of the aging process.

EXTRACELLULAR SPACES ALSO UNDERGO CHANGE

- The Extracellular Spaces between neurons in the Hippocampus and certain Cortical Areas accumulate moderate numbers of deposits known as Senile Plaques.
- Plaques develop very slowly. They are comprised of Amyloid Bodies (a protein).
- Amyloid Protein also accumulates in scattered blood vessels in the Hippocampus and Cortex.
- A significant increase in Amyloid Protein Deposits occurs in Alzheimer's Disease.
- Researchers are uncertain about the significance of these plaques in the Healthy Elderly.

DNA DAMAGE AND THE AGING PROCESS

- Most theories hold that the body ages as a result of DNA Damage in the cells.
- The Enzymatic Machinery designed to repair faulty DNA in cellular structures becomes less efficient in late life (70s+).
- The DNA in Mitochondria (the energy provider of the cells) slowly become defective.

OXIDATION OF CELLS AND THE AGING PROCESS

- The levels of Oxidized Proteins in human cells also progressively increase with age.
- Cells from young adults with Progeria (a syndrome of premature aging) contain levels of Oxidized Proteins that approach those found in healthy 80-year-olds.
- There is evidence to suggest that such Oxidation of Proteins may lead to a loss of Mental Function.

DEGENERATION OF MYELIN AND THE AGING PROCESS

- Age-related changes occur in the Myelin that Sheathes and Insulates Axons.
- Alterations of Myelin can have a measurable effect on the speed and efficiency with which Neurons propagate electrical impulses.
- Functionally, this may translate into decreased speed of movement and thought.

THE HEALTHY ELDERLY

- In the Healthy Elderly, the extent of anatomic and physiologic alterations tends to be modest.
- The Degree of Neuronal Loss, the Accumulation of Plaques, and the Deficiency of Enzymatic Activity range from 5% to 30% above those of young adults.
- Such gradual declines often appear to have little functional effect on mental faculties.
- PET Scans of the Healthy Elderly have found that the brains of healthy people in their 80s were almost as active as people in their 20s.
- The brain appears to have considerable physiologic reserve and the ability to tolerate small losses of neuronal function.
- Epidemiological studies show that, as a group, almost 90% of all people older than 65 years are free of dementia. Fewer than 5% of people aged 65 to 75 years exhibit symptoms of Dementia.
- In people aged 75 to 84 years, 20% of people are found to exhibit some symptoms of Dementia. The percentage then jumps to 50% in people older than 85.
- Studies have found that when people in their 70s and 80s remain in good health, they show only a subtle decline in performance on tests of Memory, Perception, and Language.
- The Speed of Processing, however, significantly declines.
- While the Healthy Elderly may be unable to retrieve certain details of events or information (eg, dates, places, names) as quickly, they are nevertheless often able to recall the information minutes or hours later.
- Given enough time and a stress-reduced environment, most healthy elderly score about as well as young or middle-aged adults on tests of mental performance.
- The more complex a task is (eg, a multiple step math problem), however, the more likely a healthy elder will perform less well than a young adult.
- While one may not learn or remember quite as rapidly during healthy late life, the ability to learn and remember declines only minimally.

PROTECTING THE AGING BRAIN

- Researchers suggest that remaining physically fit and mentally active may lessen age-related cognitive deficits. Older people who consistently exercise and read perform better on cognitive tests than do sedentary individuals of the same age and health status.
- Physicians also must be cautious about the type and mix of medications they prescribe for their elderly patients.
  - People over 60 are particularly sensitive to Benzodiazepines (sedatives like Valium) and other depressants and stimulants of the central nervous system.
  - Compared with young adults, older people experience a greater decline in reasoning while such drugs are in their system.
  - The elderly are affected longer and react to lower doses more strongly.
  - Drug interactions can also have serious side effects that impair mental function.
- The same is true of Anesthesia used during surgical procedures. General anesthesia has a greater effect and lasts longer in the elderly, causing disorientation and confusion.
- Health care professionals must also be aware that Nutritional Deficiencies can mask as Dementia. Greater public awareness of Health and Nutrition in the Aging Population must be made available and readily accessible to this population and to their caregivers.

Sex Differences in Male and Female Brains

RESEARCH REGARDING SEX-BASED BRAIN DIFFERENCES

- Findings regarding the differences between Male and Female Brains have largely come from:
  - PET, SPECT, and MRI Scans
  - Research regarding Non-Injured and Brain-Injured Patients
  - Research regarding Sex-Based Hormonal Differences in Fetal Development
- It is important—as with any new research that is in the nascent stages of development—that findings are not misinterpreted and that false generalizations are NOT extrapolated from the data.

OVERLAP BETWEEN THE MALE AND FEMALE BRAIN

- Despite male and female brain differences found through research, there is still a greater overlap of brain similarity than dissimilarity.
- In both males and females, there is a substantial range of abilities within each sex.
- So that while males as a group may perform better on spatial skills than females, there are females who perform equally or better than men on spatial skill tasks, and vice versa.

THE EFFECTS OF SEX HORMONES ON BRAIN ORGANIZATION IN UTERO

- Differing patterns of ability between males and females likely reflect Different Hormonal Influences upon Fetal Development of each sex.
- The action of Estrogen (female hormone) and Androgens (male hormones—chief of which is Testosterone) establish Sexual Differentiation in utero.
- In all mammals—including Humans—embryos have the potential to develop into a male or female in the first weeks of life.
- Humans have 46 Chromosomes—XX (female) and XY (male).
- If a Y Chromosome is present, testes or male gonads form. This is the critical first step toward becoming male.
- If gonads do not produce male hormones—or if the hormones cannot act on the fetal tissue—the XY embryo defaults into the form of a female organism. This is referred to as the FEMALE DEFAULT SYSTEM.
- There are many abnormalities that can occur regarding the release of specific hormones necessary to differentiate the unisex embryo into a male. The Female Default System is an evolutionary safeguard to maintain the life of the fetus.
- If testes are formed, they produce substances that facilitate the development of a male. One such substance, Testosterone, causes Masculinization by promoting the male or Wolffian Ducts to convert to the external appearance of a scrotum and penis. Wolffian Ducts in the embryo develop into the Vas Deferens and Seminal Vesicles in the male.
- The Mullerian Ducts cause the female gonads to regress physically. The Mullerian Ducts in the embryo develop into the Uterus, Fallopian Tubes, and Vagina in a female embryo.

Abnormalities that Occur In Utero

Turner Syndrome

- Chromosomal anomaly of 45 XO. The female is missing one sex chromosome.
- Causes the absence of ovaries. The female is infertile.
- Other signs and symptoms include short stature and webbed neck.
- Females require sex hormone treatment in puberty to induce secondary sex characteristics and menstruation.

Klinefelter's Syndrome

- Chromosomal anomaly in males. The male has an extra sex chromosome: 47 XXY.
- Signs and Symptoms include small penis, small testes, sterility (no sperm production), tall and thin body type and male gynecomastia (development of male breasts). Secondary male sex characteristics are often weakly developed and do not respond well to hormonal therapy.

DHT Deficiency Syndrome

- Di-Hydro-Testosterone, also called Deficiency of 5-Alpha Reductase or DHT Deficiency Syndrome.
- Dihydrotestosterone (DHT) is a powerful Androgen converted from Testosterone (through the enzymatic action of 5-Alpha Reductase).
- This directs the Differentiation of the Penis and Scrotum.
- DHT is converted from Testosterone.
- But in DHT Deficiency Syndrome, DHT cannot be produced.
- DHT is necessary for the development of male external sexual anatomy.
- The fetus possesses the normal 46 XY Chromosomes.
- But the External Anatomy remains Female because Testosterone cannot convert into DHT, which directs the Differentiation of the Penis and Scrotum.
- The Testes develop in the 2nd month of pregnancy but do not descend.
- Mullerian Inhibiting Hormone is present and negates the development of the fallopian tubes, uterus, and vagina.
- At puberty, the common Testosterone Surge (occurring at this life stage) may trigger the transformation of a clitoral-like phallus into a small penis.
- Weak secondary male sexual characteristics may develop at this time.
- It appears that the little girl, at puberty, turns into a male.
- This anomaly has been reported commonly in the Dominican Republic.

Testicular Feminizing Syndrome

- Also called Androgen Insensitivity Syndrome.
- The embryo possesses the normal 46 XY Chromosomes.
- But the cells of the embryo are not able to respond to Testosterone.
- The internal anatomy is male but is incomplete.
- The Female Default Plan is attempted by the organism but cannot be carried out fully. Because of the presence of Testosterone, the development of Internal Female Anatomy is prevented—the development of ovaries cannot occur and the individual is infertile.
- The External Anatomy appears as a normal female, however.

Retained Mullerian Syndrome

- This disorder results in a type of pseudo hermaphrodite.
- Internal Female Sex Organs are present.
- External Male Sex Organs are also present.

- Like the Female Default System, there is an evolutionary safeguard that maintains fetal life if a prenatal anomaly prevents the XX organism from developing into a female. This is referred to as the ADAM PLAN, and it involves a Defeminization Process of the Female Template. The Adam Plan stops the embryonic development of the female sex anatomy and instead, facilitates the Masculinization of embryonic structures through Mullerian Inhibiting Hormone, Dihydrotestosterone, and Testosterone.

In Addition to Differentiating Male/Female Sex Organs, Sex Hormones May Also Organize Male/Female Behavior in Animals and Humans

Animal Studies

- If a rodent with functional male genitals is deprived of Androgens immediately after birth, male sexual behavior—such as mounting—will be reduced.
- Instead, female behaviors—such as Lordosis (arching of the back to receive intercourse)—will be increased.
- Similarly, if Androgens are administered to female rodents directly after birth, the female rodent displays more male sexual behavior and less female behavior.

Congenital Adrenal Hyperplasia in Humans

- CAH occurs as a result of a genetic anomaly that causes large amounts of Adrenal Androgens to be produced in the XX embryo.
- These are females who were exposed to an excess of Androgens in the Prenatal Stage.
- Although the consequent masculinization of genitals can be surgically corrected in early life, and drug therapy can stop the overproduction of Androgens, the effects of Prenatal Exposure on the Brain cannot be reversed.
- Females who were exposed to Prenatal Androgens grow up to be self-reported tomboys and prefer typically masculine toys more than their unaffected biological sisters.
- Researchers also found that the CAH females performed better than their unaffected biological sisters on spatial manipulation tasks.

Nature Vs. Nuture

- Researchers have suggested that the effects of Sex Hormones on Brain Organization occur so early in Development that from birth, the environment is acting on differently wired male and female brains.
- One study found that 3-year-old boys perform better at physical targeting skills (guiding or intercepting projectiles) than girls of the same age.
- Other studies have shown that the extent of experience playing sports does not account for sex differences in targeting skills exhibited by young adults.
- Some researchers have argued that sex differences in spatial rotation performance are present before puberty.

Skills that Males Perform Better Than Females

- Skills that adult males perform better than adult females include the following. It should be noted that this finding considers men as a group, without considering differences occurring in both sexes.
  - Spatial Tasks requiring the individual to Imagine Rotating an Object or Manipulating it in some way Visually.
  - Mathematical Reasoning.
  - Navigating their way through a Route.
  - Target-Directed Motor Skills (guiding or intercepting projectiles).

Skills that Females Perform Better Than Males

- Skills that adult females perform better than adult males include the following. It should be noted that this finding considers women as a group, without considering differences occurring in both sexes.
  - Rapidly identifying Matching Items (called Perceptual Speed).
  - Verbal Fluency (including the ability to find words that begin with a specific letter or fulfill some other constraint).
  - Arithmetic Calculation.
  - Recalling Landmarks from a Route.
  - Precision Fine Motor Skills.

ROUTE NAVIGATION

- In studies, men learned a route in fewer trials than women and made fewer errors.
- Once learned, however, women remembered more of the landmarks than did men.
- Women may use landmarks as a strategy to orient themselves.
- The way that men orient themselves has not as yet been established, although it is suggested that they may use spatial skills.

THE RECALL OF OBJECTS AND THEIR LOCATIONS

- In studies that tested the ability of men and women to recall objects and their locations in a confined space (a room or tabletop), it was found that women were better able to remember whether an item had been displaced or not.
- Women were also better able to replace objects in their original locations (when moved).

EVOLUTIONARY SIGNIFICANCE

- Some theorists suggest that the above sex-based differences have evolutionary roots.
- Such theorists suggest that for thousands of years during which human brain characteristics evolved, humans lived in relatively small hunter-gatherer groups.
- The division of labor between the sexes in such societies was probably distinct—as it is in existing hunter-gatherer societies.
- Men were believed to be responsible for large game hunting requiring long-distance travel.
- Women were believed to gather food near the campsite, prepare food and clothing, and care for children.
- Theorists argue that such specializations placed different evolutionary selection pressures on men and women.
- Men would require long-distance route-finding abilities to recognize a geographic area from varying locations. They would also need targeting skills for hunting.
- Women would require short-range navigation using landmarks, fine motor capabilities carried out within a circumscribed space, and perceptual discrimination skills sensitive to small changes in the environment or in children's appearance and behavior.

PET SCANS AND THE CORPUS CALLOSUM

- PET Scans have shown that the Corpus Callosum is more extensive in women than in men. Women have 11% more Neurons in the Corpus Callosum, thus creating more Neural Fibers that Connect the two Hemispheres.
- This may suggest that women are more able to integrate both Hemispheres in activities than are men.
- In Language Skills, women activate both Hemispheres to interpret the Literal Meaning of the Words (Left Brain Function) as well as the Emotional Tone of the Words (Right Brain Function).
- Men activate only the Left Hemisphere in Language Skills and have greater difficulty interpreting the subtle Emotional Tone of a conversation.
- The incidence of Language Disorders (aphasia) is higher in men than in women after damage to the Left Hemisphere.
- Research has also shown that females typically recover from Aphasia after cerebrovascular accident more readily than do men. This may relate to a female's greater ability to integrate both Hemispheres during Language Tasks.

PET SCANS AND THE LIMBIC SYSTEM

- PET Scans have also shown that men exhibit greater brain activity in the more ancient regions of the Limbic System—the regions that mediate action in response to emotion.
- Women exhibit greater activity in the phylogenetically newer and more complex regions of the Limbic System that mediate symbolic action in response to emotion.
- One researcher explained that, if a dog is angry and jumps and bites, that is an active response to the emotion of anger. If the dog bears its teeth and growls, that is a symbolic response to the emotion of anger.
- The implication may be that women are more predisposed to a symbolic display of their emotions, while men react with action to their emotions.

- Women are also more adept at recognizing symbolic expressions of emotions in others than are men. Men seem to need an overt reaction that clearly demonstrates emotion—otherwise they miss emotions that are symbolically or subtly displayed.
- Studies have shown that women are more sensitive to recognizing subtle facial expressions than are men (a Limbic System Function).
- In studies, both men and women were equally adept at recognizing happiness in the photos of faces of both men and women.
- But women were more adept at determining if a man or woman was sad. A woman's face had to be tearful for men to recognize the emotion of sadness. Subtle expressions of sadness were not recognized.

REFERENCES

SECTION 1: DIRECTIONAL TERMINOLOGY; PLANES OF THE BRAIN

Dirckx JH, ed. *Stedman's Concise Medical Dictionary for the Health Professions*. 3rd ed. Baltimore, Md: Williams & Wilkins, 1997.

Thomas, CL, ed. **Taber's Cyclopedic Medical Dictionary.** 16th ed. Philadelphia, Pa: FA Davis; 1989.

SECTION 2: DIVISION OF THE NERVOUS SYSTEM

Brodal P. *The Central Nervous System*. 5th ed. New York, NY: Oxford University Press; 1992.

Partridge LD, Partridge LD. *The Nervous System: Its Function and Its Interaction with the World*. Cambridge, Mass: MIT Press; 1993.

SECTION 3: GROSS CEREBRAL STRUCTURES

Augur AM, Lee MJ. *Grant's Atlas of Anatomy*. 10th ed. Baltimore, Md: Lippincott Williams & Wilkins; 1999.

Austin JH. Arousal pathways in the reticular formation and beyond. In: *Zen and the Brain*. Cambridge, Mass: MIT Press; 1999:159-163.

Austin JH. Remembrances and the hippocampus. In: *Zen and the Brain*. Cambridge, Mass: MIT Press; 1999:180-188.

Austin JH. The amygdala and fear. In: *Zen and the Brain*. Cambridge, Mass: MIT Press; 1999:175-179.

Austin JH. The attachments of the cingulate gyrus. In: *Zen and the Brain*. Cambridge, Mass: MIT Press; 1999:172-174.

Austin JH. The septum and pleasure. In: *Zen and the Brain*. Cambridge, Mass: MIT Press; 1999:169-171.

Austin JH. The thalamus. In: *Zen and the Brain*. Cambridge, Mass: MIT Press; 1999:263-266.

Austin JH. Visceral drives and the hypothalamus. In: *Zen and the Brain*. Cambridge, Mass: MIT Press; 1999:189-196.

Braak H, Braak E. The hypothalamus of the human adult: chiasmatic region. *Anatomical Embryology*. 1987;175:315-330.

Broak H, et al. Functional anatomy of human hippocampal formation and related structures. *Journal of Child Neurology*. 1996;11:265-275.

Dawn I, Ackermann H. Cerebellar contributions to cognition. *Behavioral Brain Research*. 1995;67:201-210.

Haines DE. *Neuroanatomy: An Atlas of Structures, Sections, and Systems*. 5th ed. Philadelphia, Pa: Lippincott Williams & Wilkins; 1999.

Hanaway J, et al. Localization of the pyramidal tract in the internal capsule. *Neurology*. 1981;31:365-366.

Hirai T, Jones EG. A new parcellation of the human thalamus on the basis of histochemical staining. *Brain Research Review*. 1989;14:1-34.

Houk JC, Wise SP. Distributed modular architectures linking basal ganglia, cerebellum, and cerebral cortex: their role in planning and controlling action. *Cerebral Cortex*. 1995;2:95-110.

Jennes L, Traurig HH, Conn PM. *Atlas of the Human Brain*. Philadelphia, Pa: Lippincott; 1995.

Kordon C. Neural mechanisms involved in pituitary control. *Neurochemistry International*. 1985;7:917-925.

Leiner HC, et al. The role of the cerebellum in the human brain. *Trends in Neurosciences*. 1993;16:453-454.

Mark LP, et al. Hippocampal anatomy and pathologic alterations on conventional MR images. *American Journal of Neuroradiology*. 1993;14:1237-1240.

Mark LP, et al. Limbic connections. *American Journal of Neuroradiology*. 1995;16:1303-1306.

Mark LP, et al. Limbic system anatomy: an overview. *American Journal of Neuroradiology*. 1993;14:349-352.

Mark LP, et al. The fornix. *American Journal of Neuroradiology*. 1993;14:1355-1358.

Mark LP, et al. The hippocampus. *American Journal of Neuroradiology*. 1993;14:709-712.

Mark LP, et al. The septal area. *American Journal of Neuroradiology*. 1994;15:273-276.

Mello Luiz EA, Villares J. Neuroanatomy of the basal ganglia. *Psychiatric Clinic of North America*. 1997;20:691-703.

Netter FH. *Atlas of Human Anatomy*. 2nd ed. East Hanover, NJ: Novartis; 1997.

Netter FH. *The Ciba Collection of Medical Illustrations. Vol 1. The Nervous System. Part 1: Anatomy and Physiology*. West Caldwell, NJ: Ciba-Geigy; 1986.

Olson TR. *ADAM: Student Atlas of Anatomy*. Baltimore, Md: Williams & Wilkins; 1996.

Posner MI, Petersen SE. The attention system of the human brain. *Annual Review of Neuroscience*. 1990;13:25-42.

Robbins TW, Everitt BJ. Arousal systems and attention. In: Gazzaniga MS, ed. *The Cognitive Neurosciences*. Cambridge, Mass: MIT Press; 1995:703-720.

Roland PE. Partition of the human cerebellum in sensory-motor activities, learning, and cognition. *Canadian Journal of Neurological Science*. 1993;20(suppl 3):S75-S77.

Shatz CJ. Dividing up the neocortex. *Science*. 1992;258:237-238.

Swanson LW, Petrovich GD. What is the amygdala? *Trends in Neurosciences*. 1998;21:323-331.

Thach WT, Goodkin HP, Keating JG. The cerebellum and the adaptive coordination of movement. *Annual Review of Neuroscience*. 1992;15:403-442.

Van Hoesen GW. Anatomy of the medial temporal lobe. *Magnetic Resonance Imaging*. 1995;13:1047-1055.

Williams PL. *Gray's Anatomy*. 38th ed. New York, NY: Churchill Livingstone; 1995.

SECTION 4: THE VENTRICULAR SYSTEM

Cutler R, Spertell R. Cerebrospinal fluid: a selective review. *Annals of Neurology*. 1982;11:1-10.

Leech RW. Normal anatomy of ventricles, meninges, subarachnoid space, and venous system. In: Leech RW, Brumback RA, eds. *Hydrocephalus: Current Clinical Concepts*. St. Louis, Mo: Mosby-Year Book; 1991:18.

Leech RW. Normal physiology of cerebrospinal fluid. In: Leech RW, Brumback RA, eds. *Hydrocephalus: Current Clinical Concepts*. St. Louis, Mo: Mosby-Year Book; 1991:30.

Leech RW, Goldstein E. Hydrocephalus: classification and mechanisms. In: Leech RW, Brumback RA, eds. *Hydrocephalus: Current Clinical Concepts*. St. Louis, Mo: Mosby-Year Book; 1991:45-70.

Nolte J. Ventricles and cerebrospinal fluid; blood supply of the brain. In: *The Human Brain: An Introduction to Its Functional Anatomy*. 3rd ed. St. Louis, Mo: Mosby-Year Book; 1993:48-92.

Norman MG, et al. Hydrocephalus. In: Norman MG, et al., eds. *Congenital Malformations of the Brain: Pathological, Embryological, Clinical, Radiological, and Genetic Aspects*. New York, NY: Oxford University Press; 1995:333-339.

Sage MR, Wilson AJ. The blood-brain barrier: an important concept in neuroimaging. *American Journal of Neuroradiology*. 1994;15:601-622.

Trend P, et al. *Neurology*. 2nd ed. New York, NY: Churchill Livingstone; 1998.

Williams MA, Razumovsky AY. Cerebrospinal fluid circulation, cerebral edema, and intracranial pressure. *Current Opinion in Neurobiology*. 1996;6:847-853.

SECTION 5: THE CRANIUM

Augur AM, Lee MJ. *Grant's Atlas of Anatomy*. 10th ed. Baltimore, Md: Lippincott Williams & Wilkins; 1999.

Netter FH. *Atlas of Human Anatomy*. 2nd ed. East Hanover, NJ: Novartis; 1997.

Olson TR. *ADAM: Student Atlas of Anatomy*. Baltimore, Md: Williams & Wilkins; 1996.

Williams PL. *Gray's Anatomy*. 38th ed. New York, NY: Churchill Livingstone; 1995.

SECTION 6: THE MENINGES

England MA, Wakely J. *Color Atlas of the Brain and Spinal Cord*. St. Louis, Mo: Mosby-Year Book; 1991.

Leech RW. Normal anatomy of ventricles, meninges, subarachnoid space, and venous system. In: Leech RW, Brumback RA, eds. *Hydrocephalus: Current Clinical Concepts*. St. Louis, Mo: Mosby-Year Book; 1991:18.

Netter FH. *Atlas of Human Anatomy*. 2nd ed. East Hanover, NJ: Novartis; 1997.

Sage MR, Wilson AJ. The blood-brain barrier: an important concept in neuroimaging. *American Journal of Neuroradiology*. 1994;15:601-622.

Section 7: Spinal Cord Anatomy

Adams RD, et al. *Principles of Neurology*. 6th ed. New York, NY: McGraw-Hill; 1997.

Adams RW, et al. The distribution of muscle weakness in upper motor neuron lesions affecting the lower limb. *Brain*. 1990;113:1459-1476.

Brown AG. *Organization of the Spinal Cord: The Anatomy and Physiology of Identified Neurons*. Berlin: Springer-Verlag; 1981.

Davidoff RA. Skeletal muscle tone and the misunderstood stretch reflex. *Neurology*. 1992;42:951-963.

Davidoff RA, Hackman JC. Aspects of spinal cord structure and reflex function. *Clinical Neurology*. 1991;9:533-550.

England MA, Wakely J. *Color Atlas of the Brain and Spinal Cord*. St. Louis, Mo: Mosby-Year Book; 1991.

Gilman SG, et al. *Manter and Gatz's Essentials of Clinical Neuroanatomy and Neurophysiology*. 9th ed. Philadelphia, Pa: FA Davis; 1996.

Priestly JV. Neuroanatomy of the spinal cord: current research and future prospects. *Paraplegia*. 1987;25:198-204.

Schoenen J. Dendritic organization of the human spinal cord: the motorneurons. *Journal of Comprehensive Neurology*. 1982;211:226-247.

Schoenen J. The dendritic organization of the human spinal cord: the dorsal horn. *Neuroscience*. 1982;7:2057-2087.

Storrs BB, et al. The tethered cord syndrome. *International Pediatrics*. 1990;5:99-103.

Section 8: Cranial Nerves

Deleu D, et al. Vertical one-and-a half syndrome: supranuclear downgaze paralysis with monocular elevation palsy. *Archives of Neurology*. 1989;46:1361-1363.

Geraint F. *Neurological Examination Made Easy*. New York, NY: Churchill Livingstone; 1993.

Gilman SG, et al. *Manter and Gatz's Essentials of Clinical Neuroanatomy and Neurophysiology*. 9th ed. Philadelphia, Pa: FA Davis; 1996.

Hommel M, Bogousslavsky J. The spectrum of vertical gaze palsy following unilateral brainstem stroke. *Neurology*. 1991;41:1229-1234.

Liu GT, et al. Unilateral oculomotor palsy and bilateral ptosis from paramedian midbrain infarction. *Archives of Neurology*. 1991;48:983-986.

Marshall RS, et al. Dissociated vertical nystagmus and internuclear ophthalmoplegia from a midbrain infarction. *Archives of Neurology*. 1991;1034-1305.

Mauskop A. Trigeminal neuralgia. *Journal of Pain Symptom Management*. 1993;8:148-154.

Miller AJ. Neurophysiological basis of swallowing. *Dysphagia*. 1986;1:91-100.

Rappaport AH, Devor M. Trigeminal neuralgia: the role of self-sustaining discharge in the trigeminal ganglion. *Pain*. 1994;56:127-138.

Sengpiel F, Blakemore C. *The neural basis of suppression and amblyopia in strabismus, part 2*. Eye. 1996;10:250-258.

Trend P, et al. *Neurology*. 2nd ed. New York, NY: Churchill Livingstone; 1998.

Waespe W, Wichman W. Oculomotor disturbances during visual vestibular interactions in Wallenberg's lateral medullary syndrome. *Brain*. 1990;113:821-846.

Wilson-Pauwels L, Akesson E, Steward P. *Cranial Nerves: Anatomy and Clinical Comments*. Toronto: BC Decker; 1988.

Section 9: Sensory Receptors

Bear MF, Connors BW, Paradiso MA. *Neuroscience: Exploring the Brain*. Baltimore, Md: Williams & Wilkins; 1996.

Carlson NR. *Physiology of Behavior*. 5th ed. Boston, Mass: Allyn and Bacon; 1994.

Hillman H, Darman J. *Atlas of the Cellular Structure of the Human Nervous System*. London: Academic Press; 1991.

Netter FH. *The Ciba Collection of Medical Illustrations*. Vol 1. *The Nervous System. Part 1: Anatomy and Physiology*. West Caldwell, NJ: Ciba-Geigy; 1986.

Nicholls JG, Martin AR, Wallace B. *From Neuron to Brain: A Cellular Approach to the Function of the Nervous System.* 3rd ed. Saunderland, Mass: Sinauer; 1993.

Partridge LD, Partridge LD. *The Nervous System: Its Function and Its Interaction with the World.* Cambridge, Mass: MIT Press; 1993.

Shepherd G. *Neurobiology.* 3rd ed. New York, NY: Oxford University Press; 1994.

Vallbo AB. Single-afferent neurons and somatic sensation in humans. In: Gazzaniga MS, ed. *The Cognitive Neurosciences.* Cambridge, Mass: MIT Press; 1995:237-251.

Vallbo AB, Johansson RS. Properties of cutaneous mechanoreceptors in the human hand related to touch sensation. *Human Neurobiology.* 1984;3:3-14.

Section 10: Neurons and Action Potentials

Bergman RA, et al. *Histology.* Philadelphia, Pa: Saunders; 1996.

Dowling JE. The chemistry of synaptic transmission. In: *Neurons and Networks: An Introduction to Neuroscience.* Cambridge, Mass: Belknap Press of Harvard University Press; 1992:125-150.

Fawcett D, et al. *A Textbook of Histology.* 12th ed. New York, NY: Chapman & Hall; 1994.

Hall ZW. *An Introduction to Molecular Neurobiology.* Sunderland, Mass: Sinauer; 1992.

Hillman H, Darman J. *Atlas of the Cellular Structure of the Human Nervous System.* London: Academic Press; 1991.

Huguenard J, McCormick D. *Electrophysiology of the Neuron.* New York, NY: Oxford University Press; 1994.

Jessel T, Kandel E. Synaptic transmission: a bidirectional and self-modifiable form of cell-cell communication. *Cell.* 1993;72:1-30.

Kelly R. Storage and release of neurotransmitters. *Cell.* 1993;72:43-53.

Levitan I, Kaczmarek L. *The Neuron: Cell and Molecular Biology.* New York, NY: Oxford University Press; 1991.

Meredith GE, Arbuthnott GW. *Morphological Investigations of Single Neurons in Vitro.* New York, NY: John Wiley; 1993.

Nicholls JG, Martin AR, Wallace B. *From Neuron to Brain: A Cellular Approach to the Function of the Nervous System.* 3rd ed. Saunderland, Mass: Sinauer; 1993.

Shepherd G. *Foundations of the Neuron Doctrine.* New York, NY: Oxford University Press; 1991.

Shepherd G. *Neurobiology.* 3rd ed. New York, NY: Oxford University Press; 1994.

Shepherd G. *The Synaptic Organization of the Brain.* 4th ed. New York, NY: Oxford University Press; 1990.

Unwin N. Neurotransmitter action: opening of ligand-gated ion channels. *Cell.* 1993;72:31-41.

Wu L, Saggu P. Presynaptic inhibition of elicited neurotransmitter release. *Trends in Neurosciences.* 1997;20:204-212.

Zufall F. Cyclic nucleotide-gated channels, nitric oxide, and neural function. *Neuroscientist.* 1996;2:24-32.

Section 11: Special Sense Receptors

Ashmore JK, Kolston PJ. Hair cell based amplification in the cochlea. *Current Opinion in Neurobiology.* 1994;4:503-508.

Ayres AJ. *Sensory Integration and Praxis Tests.* Los Angeles, Calif: Western Psychological Services.

Buser P, Imbert M. *Audition.* Cambridge, Mass: MIT Press; 1992.

Dallos P. The active cochlea. *Journal of Neuroscience.* 1992;12:4575-4585.

Dallos P, Popper AN, Fay RR, eds. *The Cochlea.* New York, NY: Springer-Verlag; 1996.

Engen T. *Odor Sensation and Memory.* New York, NY: Praeger; 1991.

Gilbertson TA. The physiology of vertebrate taste reception. *Current Opinion in Neurobiology.* 1993;3:532-539.

Hudspeth AJ. The hair cells of the inner ear. *Scientific American.* 1983;248:54-64.

Hudspeth AJ, Markin V. The ear's gears. *Physics Today.* 1994;47:22-28.

Kinnamon SC, Cummings TA. Chemosensory transduction mechanisms in taste. *Annual Review of Physiology.* 1992;54:715-731.

Lancet D. Exclusive receptors. *Nature.* 1994;372:321-322.

Lekwuwa GU, Barnes GR. Cerebral control of eye movements: the relationship between cerebral lesion sites and smooth pursuit deficits. *Brain.* 1996;119:473-490.

Lim DJ. Functional structure of the organ of corti: a review. *Hearing Research*. 1986;22:117-146.

McDevitt D. *Cell Biology of the Eye*. New York, NY: Academic Press; 1982.

McLaughlin S, Margolskee RF. The sense of taste. *American Scientist*. 1994;82:538-545.

Newman NJ. Neuro-ophthalmology: the afferent visual system. *Current Opinion in Neurobiology*. 1993;6:738-746.

Nieuwenhuys R. Anatomy of the auditory pathways with emphasis on the brainstem. *Advanced Otorhinolaryngology*. 1984;34:25-38.

Ressler J, Sullivan SL, Buck LB. A molecular dissection of spatial patterning in the olfactory system. *Current Opinion in Neurobiology*. 1994;4:588-596.

Shepherd GM. Synaptic organization of the mammalian olfactory bulb. *Physiology Review*. 1972;52:864-917.

Urasaki E, et al. Cortical tongue area studied by chronically implanted subdural electrodes: with special reference to parietal motor and frontal sensory responses. *Brain*. 1994;117:117-132.

Zeki S. The visual image in mind and brain. *Scientific American*. 1992;267:69-76.

SECTION 12: VESTIBULAR SYSTEM

Blum PS, et al. Thalamic components of the ascending vestibular system. *Experimental Neurology*. 1979;64:587-603.

Brandt T. Man in motion: historical and clinical aspects of vestibular function. *Brain*. 1991;114:2159-2174.

Herdman SJ, ed. *Vestibular Rehabilitation*. Philadelphia, Pa: FA Davis; 1994.

Horak FB, Shupert CL. Role of the vestibular system in postural control. In: Herdman SJ, ed. *Vestibular Rehabilitation*. Philadelphia, Pa: FA Davis; 1994:22-46.

Leigh RJ. Human vestibular cortex. *Annals of Neurology*. 1994;35:383-384.

SECTION 13: AUTONOMIC NERVOUS SYSTEM

Appenzeller O. *The Autonomic Nervous System: An Introduction to Basic Clinical Concepts*. New York, NY: Elsevier; 1990.

Barajas-Lopez C, Huizinga JD. New transmitters and new targets in the autonomic nervous system. *Current Opinion in Neurobiology*. 1993;3:1020-1027.

Gilman SG, et al. *Manter and Gatz's Essentials of Clinical Neuroanatomy and Neurophysiology*. 9th ed. Philadelphia, Pa: FA Davis; 1996.

Janig W, McLachlan EM. Characteristics of function-specific pathways in the sympathetic nervous system. *Trends in Neurosciences*. 1992;15:475-481.

Kingsley RE. *The autonomic nervous system. In: Concise Text of Neuroscience*. 2nd ed. Philadelphia, Pa: Lippincott Williams & Wilkins; 2000:471-487.

SECTION 14: ENTERIC NERVOUS SYSTEM

De-Giorgio, et al. Primary enteric neuropathies underlying gastrointestinal motor dysfunction. *Scandinavian Journal of Gastroenterology*. 2000;35:114-122.

Ekblad E, Mei Q, Sundler F. Innervation of the gastric mucosa. *Microscopy Research and Technique*. 2000;48:241-257.

Neunlist M, Peters S, Schemann M. Multisite optical recording of excitability in the enteric nervous system. *Neurogastroenterology and Motility*. 1999;11:393-404.

SECTION 15: PAIN

Austin JH. Pain and the relief of pain. In: *Zen and the Brain*. Cambridge, Mass: MIT Press; 1999:352-354.

Austin JH. The brain's own opioids. In: *Zen and the Brain*. Cambridge, Mass: MIT Press; 1999:213-222.

Bonica JJ. *The Management of Pain*. Philadelphia, Pa: Lea & Febiger; 1990.

Bracciano AG. Pain theory and perception. In: Bracciano AG, ed. *Physical Agent Modalities: Theory and Application for the Occupational Therapist*. Thorofare, NJ: SLACK Incorporated; 2000.

Bracciano AG. *Physical Agent Modalities: Theory and Application for the Occupational Therapist*. Thorofare, NJ: SLACK Incorporated; 2000.

Burckhardt CS, et al. A randomized, controlled clinical trial of education and physical training for women with fibromyalgia. *Journal of Rheumatology*. 1994;21:714-720.

Glantz RH. Neurologic evaluation of low back pain. In: Weiner WJ, Goetz CG, eds. *Neurology for the Non-Neurologist.* 4th ed. Philadelphia, Pa: Lippincott Williams & Wilkins; 1999:317-326.

Goldstein JA. *Betrayal by the Brain: The Neurologic Basis of Chronic Fatigue Syndrome, Fibromyalgia Syndrome, and Related Neural Network Disorders.* New York, NY: Haworth Medical Press; 1996.

Headley B. Chronic pain management. In: O'Sullivan SB, Schmitz TJ, eds. *Physical Rehabilitation: Assessment and Treatment.* 13th ed. Philadelphia, Pa: FA Davis; 1994:577-602.

Lipp J. Possible mechanisms of morphine analgesia. *Clinical Neuropharmacology.* 1991;14:131-147.

Melzack R, Wall PD. *The Challenges of Pain.* 2nd ed. New York, NY: Penguin; 1996.

Melzack R, Wall PD. *The Textbook of Pain.* 3rd ed. Edinburgh: Churchill Livingstone; 1994.

Sidall PJ, Cousins MJ. Neurobiology of pain. *International Anesthesiology Clinic.* 1997;35:1-26.

Takeshige C, Zhao WH, Gvo SY. Convergence from the preoptic area and arcuate nucleus to the median eminence in acupuncture and nonacupuncture point stimulation analgesia. *Brain Research Bulletin.* 1991;26:771-778.

Willis WD. Neuroanatomy of the pain system and of the pathways that modulate pain. *Journal of Clinical Neurophysiology.* 1997;14:2-31.

Wolfe F, Simons DG. The fibromyalgia and myofascial pain syndromes: a study of tender points and trigger points in persons with fibromyalgia, myofascial pain syndrome, and no disease. *Journal of Rheumatology.* 1992;19:944-951.

SECTION 16: PERIPHERAL NERVE INJURY AND REGENERATION

Adams RD, et al. *Principles of Neurology.* 6th ed. New York, NY: McGraw-Hill; 1997.

Campbell WW. Diagnosis and management of common compression and entrapment neuropathies. *Neurologic Clinic.* 1997;15:549-567.

Dawson DM, et al. *Entrapment Neuropathies.* Boston, Mass: Little Brown; 1990.

Fisher M. Peripheral neuropathy. In: Weiner WJ, Goetz CG, eds. *Neurology for the Non-Neurologist.* 4th ed. Philadelphia, Pa: Lippincott Williams & Wilkins; 1999:187-203.

Gelb DJ. *Introduction to Clinical Neurology.* Boston, Mass: Butterworth-Heinemann; 1995.

Gilman SG, et al. *Manter and Gatz's Essentials of Clinical Neuroanatomy and Neurophysiology.* 9th ed. Philadelphia, Pa: FA Davis; 1996.

Greene DA, et al. Diabetic neuropathy. *Annual Review of Medicine.* 1990;41:303-317.

Kasdan M, et al. Carpal tunnel syndrome: management techniques. *Patient Care.* 1993;27:111-112, 115, 123-126.

Kasdan M, et al. Carpal tunnel syndrome: the workup. *Patient Care.* 1993;27:97-102, 104-108.

Lundborg G. *Nerve Injury and Repair.* Philadelphia, Pa: Churchill Livingstone; 1988.

Stillwell GK, et al. Rehabilitation procedures. In: Dyck PJ, Thomas PK, Griffin JW, eds. *Peripheral Neuropathy.* Vol 2. Philadelphia, Pa: Saunders; 1993:1692-1708.

Szabo R, Steinbert D. Nerve entrapment syndromes in the wrist. *Journal of the American Academy of Orthopaedic Surgery.* 1994;2:115-123.

Terzis JK, Smith KL. *The Peripheral Nerve: Structure, Function, and Reconstruction.* New York, NY: Raven Press; 1990.

Trend P, et al. *Neurology.* 2nd ed. New York, NY: Churchill Livingstone; 1998.

Zelig G, et al. The rehabilitation of patients with severe Guillain-Barre syndrome. *Paraplegia.* 1988;26:250-254.

SECTION 17: PHANTOM LIMB PHENOMENON

Adams RD, et al. *Principles of Neurology.* 6th ed. New York, NY: McGraw-Hill; 1997.

Gelb DJ. *Introduction to Clinical Neurology.* Boston, Mass: Butterworth-Heinemann; 1995.

Gilman SG, et al. *Manter and Gatz's Essentials of Clinical Neuroanatomy and Neurophysiology.* 9th ed. Philadelphia, Pa: FA Davis; 1996.

Melzak R. Phantom limbs. *Scientific American.* 1992;266:120-126.

Sacks O. *The Man Who Mistook His Wife for a Hat and Other Clinical Tales.* New York, NY: Harper Perennial; 1990.

Trend P, et al. *Neurology.* 2nd ed. New York, NY: Churchill Livingstone; 1998.

Van Deusen J. *Body Image and Perceptual Dysfunction in Adults.* Philadelphia, Pa: Saunders; 1993.

SECTION 18: SPINAL CORD TRACTS

Davidoff RA. The dorsal columns. *Neurology*. 1989;39:1377-1385.

Davidoff RA. The pyramidal tract. *Neurology*. 1990;40:332-339.

Hanaway J, et al. Localization of the pyramidal tract in the internal capsule. *Neurology*. 1981;31:365-366.

Holstege G. Descending motor pathways and the spinal motor system. *Progressive Brain Research*. 1991;87:307-421.

Nathan PW, et al. Sensory effects in man of lesions of the posterior columns and of some other afferent pathways. *Brain*. 1986;109:1003-1041.

Nathan PW, et al. The corticospinal tract in man: course and location of fibers at different segmental levels. *Brain*. 1990;113:303-324.

Netter FH. *The Ciba Collection of Medical Illustrations. Vol 1. The Nervous System. Part 1: Anatomy and Physiology*. West Caldwell, NJ: Ciba-Geigy; 1986.

Priestley JV. Neuroanatomy of the spinal cord: current research and future prospects. *Paraplegia*. 1987;25:198-204.

Schoenen J. Dendritic organization of the human spinal cord: the dorsal horn. *Neuroscience*. 1982;7:2057-2087.

Schoenen J. Dendritic organization of the human spinal cord: the motorneurons. *Journal of Comprehensive Neurology*. 1982;211: 226-247.

SECTION 19: SPINAL CORD INJURY AND DISEASE

Adams RD, et al. *Principles of Neurology*. 6th ed. New York, NY: McGraw-Hill, 1997.

Adams RW, et al. The distribution of muscle weakness in upper motor neuron lesions affecting the lower limb. *Brain*. 1990;113:1459-1476.

Attrice MB, et al. Traumatic spinal cord injury. In: Umphred DA, ed. *Neurological Rehabilitation*. 3rd ed. St. Louis, Mo: Mosby; 1995:484-534.

Biller J, Brazis PW. The localization of lesions affecting the spinal cord. In: Brazis PW, ed. *Localization in Clinical Neurology*. Boston, Mass: Little Brown; 1985.

Damasio H, Damasio AR. *Lesion Analysis*. New York, NY: Oxford University Press; 1989.

Davidoff RA. Skeletal muscle tone and the misunderstood stretch reflex. *Neurology*. 1992;42:951-963.

Davidoff RA, Hackman JC. Aspects of spinal cord structure and reflex function. *Clinical Neurology*. 1991;9:533-550.

Ditunno JF, Formal CS. Chronic spinal cord injury. *New England Journal of Medicine*. 1994;330:550-556.

Gelb DJ. *Introduction to Clinical Neurology*. Boston, Mass: Butterworth-Heinemann; 1995.

Gilman SG, et al. *Manter and Gatz's Essentials of Clinical Neuroanatomy and Neurophysiology*. 9th ed. Philadelphia, Pa: FA Davis; 1996.

Nathan PW, et al. Sensory effects in man of lesions of the posterior columns and of some other afferent pathways. *Brain*. 1986;109:1003-1041.

Trend P, et al. *Neurology*. 2nd ed. New York, NY: Churchill Livingstone; 1998.

SECTION 20: PROPRIOCEPTION

Bear MF, Connors BW, Paradiso MA. Proprioceptors. In: *Neuroscience: Exploring the Brain*. Baltimore, Md: Williams & Wilkins; 1996:318-320.

Burke D, Gandevia SC. Muscle spindles, muscle tone and the fusimotor system. In: Gandevia SC, Burke D, Anthony M, eds. *Science and Practice in Clinical Neurology*. Cambridge, England: Cambridge University Press; 1993:89-105.

Carlson NR. Reflex control of movement. In: *Physiology of Behavior*. 5th ed. Boston, Mass: Allyn and Bacon; 1994:230-234.

Matthews GG. *Cellular Physiology of Nerve and Muscle*. 2nd ed. Boston, Mass: Blackwell; 1991.

Moberg E. The role of cutaneous afferents in position sense, kinesthesia, and motor function of the hand. *Brain*. 1983;106:1-19.

Sacks O. *The Man Who Mistook His Wife for a Hat and Other Clinical Tales*. New York, NY: Harper Perennial; 1990.

Section 21: Disorders of Muscle Tone

Adams RD, et al. *Principles of Neurology.* 6th ed. New York, NY: McGraw-Hill; 1997.

Ashby PD. The neurophysiology of human spinal spasticity. In: *Science and Practice in Clinical Neurology.* Cambridge, England: Cambridge University Press; 1993: 106-129.

Borg-Stein J, Stein J. Pharmacology of botulinum toxin and implications for use in disorders of muscle tone. *Journal of Head Trauma Rehabilitation.* 1993;8:103-106.

Campbell S, et al. The effects of intrathecally administered baclofen on function in patients with spasticity. *Physical Therapy.* 1995;75:352-362.

Cromwell SJ, Paquette VL. The effect of botulinum toxin A on the function of a person with poststroke quadriplegia. *Physical Therapy.* 1996;76:395-402.

Davidoff RA. Skeletal muscle tone and the misunderstood stretch reflex. *Neurology.* 1992;42:951-963.

Dietz V. Spasticity: exaggerated reflexes or movement disorder? In: Forssberg H, Hirshfeld H, eds. *Movement Disorders in Children.* Basal: Karger; 1992:225-233.

Dietz V, et al. Reflex activity and muscle tone during elbow movements in patients with spastic paresis. *Annals of Neurology.* 1991;30:767-779.

Factor SA, Weiner WJ. Hyperkinetic movement disorders. In: Weiner WJ, Goetz CG, eds. *Neurology for the Non-Neurologist.* 4th ed. Philadelphia, Pa: Lippincott Williams & Wilkins; 1999:143-177.

Gilman SG, et al. *Manter and Gatz's Essentials of Clinical Neuroanatomy and Neurophysiology.* 9th ed. Philadelphia, Pa: FA Davis; 1996.

Hesse S, et al. Botulinum toxin treatment for lower limb extensor spasticity in chronic hemiparetic patients. *Journal of Neurology, Neurosurgery, and Psychiatry.* 1994;57:1321-1324.

Levin M, Hui-chan C. Relief of hemiparetic spasticity by TENS is associated with improvement in reflex and voluntary motor functions. *Electroencephalograph and Clinical Neurophysiology.* 1992;85:131-142.

Penn R. Intrathecal baclofen for spasticity of spinal origin: seven years of experience. *Journal of Neurosurgery.* 1992;77:236-240.

Sahrmann S, Norton B. The relationship of voluntary movement to spasticity in the upper motor neuron syndrome. *Annals of Neurology.* 1977;2:460-465.

Trend P, et al. *Neurology.* 2nd ed. New York, NY: Churchill Livingstone; 1998.

Section 22: Motor Functions and Dysfunctions of the CNS: Cortex, Basal Ganglia, Cerebellum

Ackermann H, et al. Speech deficits in ischemic cerebellar lesions. *Journal of Neurology.* 1992;239:223-227.

Albin RL, et al. The functional anatomy of basal ganglia disorders. *Trends in Neurosciences.* 1989;12:366-375.

Alexander GE, et al. Basal ganglia-thalamocortical circuits: parallel substrates for motor, oculomotor, prefrontal, and limbic functions. *Progress in Brain Research.* 1990;85:119-146.

Awaad Y. Tics in Tourette syndrome: new treatment options. *Journal of Child Neurology.* 1999;14:316-319.

Bhatia KP, Marsden CD. The behavioral and motor consequences of facial lesions of the basal ganglia in man. *Brain.* 1994;117:859-876.

Brooks V. *The Neural Basis of Motor Control.* New York, NY: Oxford University Press; 1986.

Carr J, Shepard R. *Movement Science: Foundations for Physical Therapy Rehabilitation.* 2nd ed. Gaithersburg, Md: Aspen; 2000.

Chaves CJ, et al. Cerebellar infarcts. *Current Neurology.* 1994;14:143-177.

Collard M, Chevalier Y. Vertigo. *Current Opinion in Neurobiology.* 1994;7:88-92.

Cordo P, Harnad S. *Movement Control.* Cambridge, England: Cambridge University Press, 1994.

Daum I, Ackerman H. Cerebellar contributions to cognition. *Behavioral Brain Research.* 1995;67:201-210.

Dogali M, et al. Stereotactic ventral pallidotomy for Parkinson's disease. *Neurology.* 1995;45:753-761.

Donoghue JP, Sanes JN. Motor areas of the cerebral cortex. *Journal of Clinical Neurophysiology.* 1994;11:382-396.

Factor SA, Weiner WJ. Hyperkinetic movement disorders. In: Weiner WJ, Goetz CG, eds. *Neurology for the Non-Neurologist*. 4th ed. Philadelphia, Pa: Lippincott Williams & Wilkins; 1999:143-177.

Flahert AW, Graybiel AM. Anatomy of basal ganglia. In: Marsden CD, Fahn S, eds. *Movement Disorders*. Boston, Mass: Butterworth-Heinemann, 1994:3.

Graybiel AM. Neurotransmitters and neuromodulators in the basal ganglia. *Trends in Neurosciences*. 1990;13:244-254.

Houk JC, Wise SP. Distributed modular architectures linking basal ganglia, cerebellum, and cerebral cortex: their role in planning and controlling action. *Cerebral Cortex*. 1995;2:95-110.

Jensen JM. Vertigo and dizziness. In: Weiner WJ, Goetz, CG, eds. *Neurology for the Non-Neurologist*. 4th ed. Philadelphia, Pa: Lippincott Williams & Wilkins; 1999:205-219.

Jueptner M, et al. Localization of a cerebellar timing process using PET. *Neurology*. 1995;45:1540-1545.

Kaas JH. The reorganization of sensory and motor maps in adult mammals. In: Gazzaniga MS, ed. *The Cognitive Neurosciences*. Cambridge, Mass: MIT Press; 1995:51-72.

Karni A, et al. Functional MRI evidence for adult motor cortex plasticity during motor skill learning. *Nature*. 1995;377:155-158.

Koller WC. Chorea, hemichorea, hemiballismus, choreoathetosis and related disorders of movement. *Current Opinion in Neurology and Neurosurgery*. 1991;4:350-353.

Leiner HC, et al. The role of the cerebellum in the human brain. *Trends in Neurosciences*. 1993;16:453-454.

Lichter DG, et al. Influence of family history on clinical expression of Tourette's syndrome. *Neurology*. 1999;52:308-316.

Marsden CD. Movement disorders and the basal ganglia. *Trends in Neurosciences*. 1986;9:512-515.

Oyanagi K, et al. Quantitative investigation of the substantia nigra in Huntingdon's disease. *Annals of Neurology*. 1989;26:13-19.

Parent A. Extrinsic connections of the basal ganglia. *Trends in Neurosciences*. 1990;13:254-258.

Petersen B, et al. Reduced basal ganglia volumes in Tourette's syndrome using three-dimensional reconstruction techniques from magnetic resonance images. *Neurology*. 1993;43:941-949.

Remy P, et al. Peduncular rubral tremor and dopaminergic denervation: a PET study. *Neurology*. 1995;45:472-477.

Robertson MM, et al. The Tourette Syndrome Diagnostic Confidence Index: development and clinical associates. *Neurology*. 1999;53:2108-2112.

Roland PE. Partition of the human cerebellum in sensory-motor activities, learning and cognition. *Canadian Journal of Neurological Science*. 1993;20:S75-S77.

Rothwell J. *Control of Human Voluntary Movement*. 2nd ed. London: Chapman and Hill; 1994.

Rumeau C, et al. Localization of hand function in the sensorimotor cortex: MR and functional correlation. *American Journal of Neuroradiology*. 1994;15:567-572.

Sacks O. *Awakenings*. New York, NY: Harper Perennial; 1990.

Sacks O. *The Man Who Mistook His Wife for a Hat and Other Clinical Tales*. New York, NY: Harper Perennial; 1990.

Sharpe JA, Johnston JL. Vertigo and nystagmus. *Current Opinion in Neurology and Neurosurgery*. 1990;3:789-795.

Sheppard DM, et al. Tourette's and comorbid syndromes: obsessive compulsive and attention deficit hyperactivity disorder. A common etiology? *Clinical Psychology Review*. 1999;19:531-552.

Sherman EM, et al. Sustained attention and impulsivity in children with Tourette syndrome: comorbidity and confounds. *Journal of Clinical and Experimental Neuropsychology*. 1998;20:644-657.

Smith AD, Bolam JP. The neural network of the basal ganglia as revealed by the study of synaptic connections of identified neurons. *Trends in Neurosciences*. 1990;13:259-265.

Stoetter B, et al. Functional neuroanatomy of Tourette syndrome: limbic-motor interactions studied with FDG PET. *Advanced Neurology*. 1992;58:213-226.

Tanji J. The supplementary motor area in the cerebral cortex. *Neuroscience Research*. 1994;19:251-268.

Weiner WJ, Shulman LM. Parkinson's disease. In: Weiner WJ, Goetz CG, eds. *Neurology for the Non-Neurologist*. 4th ed. Philadelphia, Pa: Lippincott Williams & Wilkins; 1999:129-141.

Yoshida M. The neural mechanism underlying Parkinsonism and dyskinesia: differential roles of the putamen and caudate nucleus. *Neuroscience Research*. 1991;12:31-40.

Section 23: Sensory Functions and Dysfunctions of the Central Nervous System

Gelb DJ. *Introduction to Clinical Neurology*. Boston, Mass: Butterworth-Heinemann; 1995.

Gilman SG, et al. *Manter and Gatz's Essentials of Clinical Neuroanatomy and Neurophysiology*. 9th ed. Philadelphia, Pa: FA Davis; 1996.

Kaas JH. The reorganization of sensory and motor maps in adult mammals. In: Gazzaniga MS, ed. *The Cognitive Neurosciences*. Cambridge, Mass: MIT Press; 1995:51-72.

Land MF, Fernald RD. The evolution of eyes. *Annual Review of Neuroscience*. 1992;15:1-30.

Ligeois-Chauvel C, et al. Localization of the primary auditory area in man. *Brain*. 1994;114:139-153.

Luders H, et al. The second sensory area in humans: evoked potential and electrical stimulation studies. *Annals of Neurology*. 1985;17:177-184.

Merzenich MM, Jenkins WM. Reorganization of cortical representation of the hand following alterations of skin inputs. *Journal of Hand Therapy*. 1993:89-104.

Miyashita Y. Inferior temporal cortex: where visual perception meets memory. *Annual Review of Neuroscience*. 1993;16:245-264.

Rumeau C, et al. Localization of hand function in the sensorimotor cortex: MR and functional correlation. *American Journal of Neuroradiology*. 1994;15:567-572.

Sacks O. *A Leg to Stand On*. New York, NY: Harper Perennial; 1993.

Sacks O. *An Anthropologist on Mars*. New York, NY: Vintage Books; 1995.

Sacks O. *The Man Who Mistook His Wife for a Hat and Other Clinical Tales*. New York, NY: Harper Perennial; 1990.

Sutherling WW, et al. Cortical sensory representation of the human hand: size of finger regions and non-overlapping digit somatotopy. *Neurology*. 1992;42:1020-1028.

Unsworth C. *Cognitive and Perceptual Dysfunction: A Clinical Reasoning Approach to Evaluation and Dysfunction*. Philadelphia, Pa: FA Davis; 1999.

Zeki S. The visual image in mind and brain. *Scientific American*. 1992;267:69-76.

Zoltan B. *Vision, Perception, and Cognition: A Manual for the Evaluation and Treatment of the Neurologically Impaired Adult*. 3rd ed. Thorofare, NJ: SLACK Incorporated; 1996.

Section 24: Thalamus and Brainstem Sensory and Motor Roles: Function and Dysfunction

Adams RD, et al. *Principles of Neurology*. 6th ed. New York, NY: McGraw-Hill; 1997.

Bogousslavsky J, et al. Respiratory failure and unilateral caudal brainstem infarction. *Annals of Neurology*. 1990;28:668-673.

Brazius PW. The localization of lesions affecting the brainstem. In: *Localization in Clinical Neurology*. Boston, Mass: Little Brown; 1985:225-238.

Hirai T, Jones EG. A new parcellation of the human thalamus on the basis of histochemical staining. *Brain Research Review*. 1989;14:1-34.

Kinney HC, et al. Neuropathological findings in the brain of Karen Ann Quinlan: the role of the thalamus in the persistent vegetative state. *New England Journal of Medicine*. 1994;330:1469-1475.

Kotchabhakdi N, et al. Afferent projections to the thalamus from the perihypoglossal nuclei. *Brain Research*. 1980;187:457-461.

Mauguiere F, Desmedt JE. Thalamic pain syndrome of Dejerine-Roussy: differentiation of four subtypes assisted by somatosensory evoked potentials data. *Archives of Neurology*. 1988;45:1312-1320.

Topel JL, Lewis SL. Examination of the comatose patient. In: Weiner WJ, Goetz CG, eds. *Neurology for the Non-Neurologist*. 4th ed. Philadelphia, Pa: Lippincott Williams & Wilkins; 1999:59-68.

Vuilleumier P, et al. Infarction of the lower brainstem: clinical, etiological, and MRI topographical correlations. *Brain*. 1995;118:1013-1025.

Young GB, Ropper AH, Bolton CF. *Coma and Impaired Consciousness*. New York, NY: McGraw-Hill; 1997.

Section 25: Right Vs. Left Brain Functions and Disorders

Finger S, Roe D. Gustave Dax and the early history of cerebral dominance. *Archives of Neurology*. 1996;53:806-813.

Franco L, Sperry RW. Hemisphere lateralization for cognitive processing of geometry. *Neuropsychologia*. 1977;15:107-114.

Geschwind N. *Cerebral Lateralization: Biological Mechanisms, Associations, and Pathology*. Cambridge, Mass: MIT Press; 1987.

Geschwind N, Galaburda AM, eds. *Cerebral Dominance: The Biological Foundations*. Cambridge, Mass: MIT Press; 1988.

Karbe H, et al. Planum temporale and Brodmann's area 22: magnetic resonance imaging and high resolution positron emission tomography demonstrate functional left-right asymmetry. *Archives of Neurology*. 1995;52:869-874.

Phelps E, Gazzaniga M. Hemispheric differences in mneumonic processing: the effect of left hemispheric interpretation. *Neuropsychologia*. 1992;30:293-297.

Schacter SC, Devinsky O, eds. *Behavioral Neurology and the Legacy of Norman Geschwind*. Philadelphia, Pa: Lippincott Williams & Wilkins; 1997.

Schlaug G, et al. In vivo evidence of structural brain asymmetry in musicians. *Science*. 1995;267:699-701.

Sergent J. Furtive excursions into bicameral minds. *Brain*. 1990;113:537-568.

Silberman EK, Weingartner H. Hemispheric lateralization of functions related to emotion. *Brain and Cognition*. 1986;5:322-353.

Springer SP, Deutsch G. *Left Brain, Right Brain: Perspectives on Cognitive Neuroscience*. 5th ed. New York, NY: WH Freeman; 1998.

Section 26: Perceptual Functions and Dysfunctions of the Central Nervous System

Absher RR, Benson DF. Disconnection syndromes: an overview of Geschwind's contributions. *Neurology*. 1993;43:862-867.

Alexander MP, et al. Broca's area aphasia. Aphasia after lesions including the frontal operculum. *Neurology*. 1990;40:353-362.

Arnadottir G. *The Brain and Behavior: Assessing Cortical Function Through Activities of Daily Living*. St. Louis, Mo: Mosby; 1990.

Boeri R, Salmaggi A. Prosopagnosia. *Current Opinion in Neurobiology*. 1994;7:61-64.

Daffner K, Ahern G, Weintraub S. Dissociated neglect behavior following sequential strokes in the right hemisphere. *Annals of Neurology*. 1990;28:97-101.

Damasio AR. Prosopagnosia. *Trends in Neurosciences*. 1985;8:132-135.

Damasio AR. The nature of aphasias: signs and syndromes. In: Taylor Sarno M, ed. *Acquired Aphasia*. New York, NY: Academic Press; 1981:51.

Damasio AR, Damasio H. The anatomic basis of pure alexia. *Neurology*. 1983;33:1573-1583.

Damasio AR, et al. Face agnosia and the neural substrates of memory. *Annual Review of Neuroscience*. 1990;13:89-109.

De Valois RL, De Valois KK. *Spatial Vision*. New York, NY: Oxford University Press; 1988.

Feinberg TE, et al. Anosognosia and visuoverbal confabulation. *Archives of Neurology*. 1994;51:468-473.

Feinberg TE, et al. Two alien hand syndromes. *Neurology*. 1992;42:19-24.

Rizzo M, Robin DA. Simultanagnosia: a defect of sustained attention yields insights on visual information processing. *Neurology*. 1990;40:447-455.

Sacks O. *A Leg to Stand On*. New York, NY: Harper Perennial; 1993.

Sacks O. *An Anthropologist on Mars*. New York, NY: Vintage Books; 1995.

Sacks O. *The Man Who Mistook His Wife for a Hat and Other Clinical Tales*. New York, NY: Harper Perennial; 1990.

Schaffler L, et al. Comprehension deficits elicited by electrical stimulation of Broca's area. *Brain*. 1993;116:695-715.

Smith Doody R, Jankovic J. The alien hand and related signs. *Journal of Neurology, Neurosurgery, and Psychiatry*. 1992;55:806-810.

Van Deusen J. *Body Image and Perceptual Dysfunction in Adults*. Philadelphia, Pa: Saunders; 1993.

Watson RT, et al. Posterior neocortical systems subserving awareness and neglect: neglect associated with superior temporal sulcus but not area 7 lesions. *Archives of Neurology*. 1994;51:1014-1021.

Yamadori A, et al. Left unilateral agraphia and tactile anomia: disturbances seen after occlusion of the anterior cerebral artery. *Archives of Neurology*. 1980;37:88-91.

SECTION 27: BLOOD SUPPLY OF THE BRAIN: CEREBROVASCULAR DISORDERS

Adams RD, et al. *Principles of Neurology*. 6th ed. New York, NY: McGraw-Hill; 1997.

Amarenco P, et al. Infarction in the territory of the medial branch of the posterior inferior cerebellar artery. *Journal of Neurology, Neurosurgery, and Psychiatry*. 1990;53:731-735.

Amarenco P, Hauw JJ. Cerebellar infarction in the territory of the anterior and inferior cerebellar artery: a clinico-pathological study of 20 cases. *Brain*. 1990;113:139-155.

Ausman JI, et al. Vertebrobasilar insufficiency: a review. Archives of Neurology. 1985;42:803-808.

Bogousslavsky J, Regali F. Anterior cerebral artery territory infarction in the Lausanne Stroke registry: clinical and etiologic patterns. *Archives of Neurology*. 1990;47:144-150.

Damasio H. A computed tomographic guide to the identification of cerebral vascular territories. *Archives of Neurology*. 1983;40:138-142.

Fischer CM: The posterior cerebral artery syndrome. *Canadian Journal of Neurological Science*. 1986;13:232-239.

Gelb DJ. *Introduction to Clinical Neurology*. Boston, Mass: Butterworth-Heinemann; 1995.

Gilman SG, et al. *Manter and Gatz's Essentials of Clinical Neuroanatomy and Neurophysiology*. 9th ed. Philadelphia, Pa: FA Davis; 1996.

Hommel M, et al. Hemiplegia in posterior cerebral artery occlusion. *Neurology*. 1990;40:1496-1499.

Hupperts RM, et al. Infarcts in the anterior choroidal artery territory: anatomical distribution, clinical syndromes, presumed pathogenesis, and early outcome. *Brain*. 1994;117:825-834.

Kelley RE. Cerebrovascular disease. In: Weiner WJ, Goetz CG, eds. *Neurology for the Non-Neurologist*. 4th ed. Philadelphia, Pa: Lippincott Williams & Wilkins; 1999:69-84.

Marinkovic SV, et al. Perforating branches of the middle cerebral artery: microanatomy and clinical significance of their intracerebral segments. *Stroke*. 1985;16:1022-1029.

Melo TP, et al. Pure motor stroke: a reappraisal. *Neurology*. 1992;42:789-798.

Triggs WJ, Beric A. Sensory abnormalities and dysaethesias in the anterior spinal artery syndrome. *Brain*. 1992;115:189-198.

Wolfe N, et al. Frontal systems impairment following multiple lacunar infarcts. *Archives of Neurology*. 1990;47:129-132.

SECTION 28: COMMONLY USED NEURODIAGNOSTIC TESTS

Bleck TP. Clinical use of neurologic diagnostic tests. In: Weiner WJ, Goetz CG, eds. *Neurology for the Non-Neurologist*. 4th ed. Philadelphia, Pa: Lippincott Williams & Wilkins; 1999:27-38.

Kingsley RE, Jones VF. Neuroimaging techniques. In: Kingsley RE, ed. *Concise Text of Neuroscience*. 2nd ed. Philadelphia, Pa: Lippincott Williams & Wilkins; 2000:617-630.

Nayak SM, Ramsey GR. Neuroradiology—which tests to order? In: Weiner WJ, Goetz CG, eds. *Neurology for the Non-Neurologist*. 4th ed. Philadelphia, Pa: Lippincott Williams & Wilkins; 1999:39-58.

SECTION 29: NEUROTRANSMITTERS:
THE NEUROCHEMICAL BASIS OF HUMAN BEHAVIOR

American Psychiatric Association. *Diagnostic and Statistical Manual of Mental Disorders*. 4th ed. Washington, DC: American Psychiatric Association; 1994.

Berridge CW, Abercrombie ED. Relationship between locus coerulus discharge rates of norepinephrine release within neocortex as assessed by in vivo microdialysis. *Neuroscience*. 1999;93:1263-1270.

Birnbaum S, et al. A role for norepinephrine in stress-induced cognitive deficits: alpha-1-adrenoceptor mediation in the prefrontal cortex. *Biological Psychiatry*. 1999;46:1266-1274.

Clements JD. Neurotransmitter time course in the synaptic cleft: its role in central synaptic function. *Trends in Neurosciences*. 1996;19:163-170.

Cooper JR, et al. *The Biochemical Basis of Neuropharmacology*. 3rd ed. New York, NY: Oxford University Press; 1996.

Dekeyser J, et al. The mesoneocortical dopamine neuron system. *Neurology*. 1990;40:1660-1662.

Garcia NH, et al. Chronic oral L-Dopa increases dopamine and decreases serotonin excretions. *American Journal of Physiology*. 1999;277:1476-1480.

Goldstein JA. *Betrayal by the Brain: The Neurologic Basis of Chronic Fatigue Syndrome, Fibromyalgia Syndrome, and Related Neural Network Disorders*. New York, NY: Haworth Medical Press; 1996.

Johnston TG, et al. Plasma norepinephrine and prediction of outcome in major depressive disorder. *Biological Psychiatry*. 1999;46:1253-1258.

Murphy S, Grzybick D. Glial NO: normal and pathological. *Neuroscientist*. 1996;2:90-99.

Vallee RB, Bloom GS. Mechanisms of fast and slow transport. *Annual Review of Neuroscience*. 1991;14:59-92.

Wu L, Saggau P. Presynaptic inhibition of elicited neurotransmitter release. *Trends in Neurosciences*. 1997;20:204-212.

SECTION 30: THE NEUROLOGIC SUBSTRATES OF ADDICTION

American Psychiatric Association. *Diagnostic and Statistical Manual of Mental Disorders*. 4th ed. Washington, DC: American Psychiatric Association; 1994.

Buydens-Branchey L, et al. Craving for cocaine in addicted users: role of serotonergic mechanisms. *American Journal of Addictions*. 1997;6:65-73.

Comings DE, et al. Studies of the potential role of the dopamine D1 receptor gene in addictive behaviors. *Molecular Psychiatry*. 1997;2:44-56.

Greenberg BD, et al. Altered cortical excitability in obsessive-compulsive disorder. *Neurology*. 2000;54:142-147.

Horvath AT. *Sex, Drugs, Gambling, and Chocolate*. San Luis Obispo, Calif: Impact; 1998.

Kaye WH, et al. Bulimia nervosa and substance dependence: association and family transmission. *Alcohol Clinical and Experimental Research*. 1996;20:878-881.

Koob GF. Cocaine reward and dopamine receptors: love at first site. *Archives of General Psychiatry*. 1999;56:1101-1106.

Kreek MJ. Cocaine, dopamine, and the endogenous opiod system. *Journal of Addiction Disorders*. 1996;15:73-96.

Majewski MD. Cocaine addiction as a neurological disease: implications for treatment. *NIDA Research Monograph*. 1996;163:1-26.

Romach MK, et al. Attenuation of the euphoric effects of cocaine by the dopamine D1/D5 antagonist ecopipam. *Archives of General Psychiatry*. 1999;22:521-527.

Schmidt LG, et al. Serotonergic dysfunction in addiction: effects of alcohol, cigarette smoking, and heroin on platelet 5-HT content. *Psychiatry Research*. 1997;72:177-185.

Spanagel R, Weiss F. The dopamine hypothesis of reward: past and current status. *Trends in Neurosciences*. 1999;22:521-527.

Straussner SLA, Zelvin E. *Gender and Addictions: Men and Women in Treatment*. Northvale, NJ: Jason Aronsen; 1997.

Tella SR, et al. Differential reinforcing effects of cocaine and GBR-12909: biochemical evidence for divergent neuroadaptive changes in the mesolimbic dopaminergic system. *Journal of Neuroscience*. 1996;16:7416-7427.

Volkow ND, et al. Cocaine addiction: hypothesis derived from imaging studies with PET. *Journal of Addictive Diseases*. 1996;15:55-71.

Wilfley B. Binge eating among the overweight population: a serious and prevalent problem. *Journal of the American Dietetic Association*. 1996;96:58-61.

SECTION 31: LEARNING AND MEMORY

Bliss TVP, Collingridge GL. A synaptic model of memory: long-term potentiation in the hippocampus. *Nature*. 1993;361:31-39.

Gabrieli JDE. Disorders of memory in humans. *Current Opinions in Neurology and Neurosurgery.* 1993;6:93-97.

Mennemeir M, et al. Contributions to the left intraluminar and medial thalamic nuclei to memory: comparisons and report of a case. *Archives of Neurology.* 1992;49:1050-1058.

Mesulam MM. Large-scale neurocognitive networks and distributed processing for attention, language, and memory. *Annals of Neurology.* 1990;28:597-613.

Miyashita Y. Inferior temporal cortex: where visual perception meets memory. *Annual Review of Neuroscience.* 1993;16:245-263.

Moscovitch M. Models of consciousness and memory. In: Gazzaniga MS, ed. *The Cognitive Neurosciences.* Cambridge, Mass: MIT Press; 1995:1341-1356.

Perani D, et al. Evidence of multiple memory systems in the human brain: a [18F] FDG PET metabolic study. *Brain.* 1993;116:903-919.

Perry RJ, Hodges JR. Spectrum of memory dysfunction in degenerative disease. *Current Opinions in Neurobiology.* 1996;9:281-285.

Schacter D. *Searching for Memory: The Brain, the Mind, and the Past.* New York, NY: Basic Books; 1996.

Squire LR, Knowlton B, Musen G. The structure and organization of memory. *Annual Review of Psychology.* 1993;44:453-495.

Zola-Morgan S, Squire LR. Neuroanatomy of memory. *Annual Review of Neuroscience.* 1993;16:547-564.

Section 32: The Neurologic Substrates of Emotion

Adolphs R, et al. Impaired recognition of emotion in facial expressions following bilateral damage to the human amygdala. *Nature.* 1994;372:669-672.

Aggleton JP. The contribution of the amygdala to normal and abnormal emotional states. *Trends in Neurosciences.* 1993;16:328-333.

Armony JL, LeDoux JE. How the brain processes emotional information. *Annals of the National Academy of Science.* 1997;821:259-270.

Austin JH. Remembrances and the Hippocampus. In: *Zen and the Brain.* Cambridge, Mass: MIT Press; 1999:180-188.

Austin JH. The amygdala and fear. In: *Zen and the Brain.* Cambridge, Mass: MIT Press; 1999:175-179.

Austin JH. The septum and pleasure. In: *Zen and the Brain.* Cambridge, Mass: MIT Press; 1999:169-171.

Breiter HC, et al. Acute effects of cocaine on human brain activity and emotion. *Neuron.* 1997;19:591-611.

Damasio AR. *Descartes' Error: Emotion, Reason, and the Human Brain.* New York, NY: Putnam; 1994.

Davidson RJ, et al. Regional brain function, emotion, and disorders of emotion. *Current Opinion in Neurobiology.* 1999;9:228-234.

Davis M. The role of the amygdala in fear and anxiety. *Annual Review of Neuroscience.* 1992;15:353-375.

George MS, et al. Brain activity during transient sadness and happiness in healthy women. *American Journal of Psychiatry.* 1995;152:341-351.

Kalin NH. The neurobiology of fear. *Scientific American.* 1993;16:328-333.

Lang PJ, et al. Emotion, motivation, and anxiety: brain mechanisms and psychophysiology. *Biological Psychiatry.* 1998;44:1248-1263.

Maddock RJ. The retrosplenial cortex and emotion: new insights from functional neuroimaging of the human brain. *Trends in Neurosciences.* 1999;22:310-316.

Mayberg HS, et al. Reciprocal limbic-cortical function and negative mood: converging PET findings in depression and normal sadness. *American Journal of Psychiatry.* 1999;156:675-682.

Peretz I, et al. Music and emotion: perceptual determinants, immediacy, and isolation after brain damage. *Cognition.* 1998;68:111-141.

Robinson RG, Starkstein, SE. Mood disorders following stroke: new findings and future directions. *Journal of Geriatric Psychiatry.* 1989;22:1-15.

Savitzky A. Cognition, emotion, and the brain: a different view. *Medical Hypothesis.* 1999;52:357-362.

Silberman EK, Weingartner H. Hemispheric lateralization of functions related to emotion. *Brain and Cognition.* 1986;5:322-353.

Spence S, Shapiro D, Zaidel E. The role of the right hemisphere in the physiological and cognitive components of emotional processing. *Psychophysiology.* 1996;33:112-122.

SECTION 33: THE AGING BRAIN

Backman L, et al. Brain regions associated with episodic retrieval in normal aging and Alzheimer's disease. *Neurology.* 1999;52:1861-1870.

Butterfield DA, et al. Elevated oxidative stress in models of normal brain aging and Alzheimer's disease. *Life Sciences.* 1999;65:1883-1892.

Coffey CE, et al. Relation of education to brain size in normal aging: implications for the reserve hypothesis. *Neurology.* 1999;53:189-196.

Doraiswamy PM, et al. Morphometric changes of the human midbrain with normal aging: MR and stereologic findings. *American Journal of Neuroradiology.* 1992;13:383-386.

Hazzard WR, et al. *Principles of Geriatric Medicine and Gerontology.* 3rd ed. New York, NY: McGraw-Hill, 1994.

Liu J, Mori A. Stress, aging and brain oxidative damage. *Neurochemical Research.* 1999;24:1479-1497.

Long JM, et al. What counts in brain aging? Design-based stereological analysis of cell number. *Journal of Gerontology.* 1999;54:407-417.

Magnotta VA, et al. Quantitative in vivo measurement of gyrification in the human brain: changes associated with aging. *Cerebral Cortex.* 1999;9:151-160.

Nichols NR. Glial responses to steroids as markers of brain aging. *Journal of Neurobiology.* 1999;40:585-601.

Perry RJ, Hodges JR. Spectrum of memory dysfunction in degenerative disease. *Current Opinion in Neurobiology.* 1996;9:281-285.

Rapp PR, Heindel WC. Memory systems in normal and pathological aging. *Current Opinion in Neurobiology.* 1994;7:294-298.

Roberts EL. Using hippocampal slices to study how aging alters ion regulation in brain tissue. *Methods.* 1999;18:150-159.

Selkoe DJ. Aging brain, aging mind. *Scientific American.* 1992;September:135-141.

Yanker B, Mesulam MM. -Amyloid and the pathogenesis of Alzheimer's disease. *New England Journal of Medicine.* 1991;325:1849-1857.

SECTION 34: SEX DIFFERENCES IN MALE AND FEMALE BRAINS

Kimura D. Sex differences in the brain. *Scientific American.* 1992;September:119-125.

LeVay S. *The Sexual Brain.* Cambridge, Mass: MIT Press; 1993.

McEwen BS. *Permanence of brain sex differences and structural plasticity of the adult brain.* Proceedings of the National Academy of Sciences of the United States of America. 1999;96:7128-7130.

Segovia S, et al. The development of brain sex differences: a multisignaling process. *Behavioral Brain Research.* 1999;105:69-80.

Acalculia: A type of expressive aphasia that involves the inability to calculate mathematical problems.

Accommodation: The ability of the eye to focus images of near or distant objects on the retina. The ciliary muscles are responsible for changing the thickness of the lens to focus images on the retina.

Acetylcholine (ACh): A neurotransmitter that acts at the neuromuscular junction to facilitate muscle movement.

Acetylcholinesterase (AChE): The enzyme that destroys acetylcholine soon after it is released from its terminal boutons, thus terminating the postsynaptic potential.

Achromatopsia: A condition caused by a lesion to V4 that results in the loss of the memory of color. Individuals see the world in shades of gray.

Action Potential: The brief electrical impulse that provides the basis for conduction of nerve signals along the axon. It results from the brief changes in the cell's membrane permeability to sodium and potassium ions. A strong enough action potential will cause the neuron to become excited and start the conduction process.

Adenosine: An inhibitory neurotransmitter. The release of adenosine during sleep facilitates sleep maintenance.

Adiadochokinesia: Inability to perform rapid alternating movements—as in supinating and pronating one's forearms and hands quickly and synchronously.

Agnosia: Literally "not to know." Tactile Agnosia: An inability to interpret sensations through touch. Auditory Agnosia: An inability to interpret sounds. Visual Agnosia: An inability to interpret visual stimuli. All of the above agnosias result from lesions to the cortex; the sensory receptor anatomy remains intact.

Agraphesthesia: Loss of the ability to interpret letters written on the contralateral hand. Indicates damage to the secondary somatosensory area.

Agraphia: A type of expressive aphasia resulting in the inability to write intelligible words and sentences.

Agrommation: A type of expressive aphasia that involves the inability to arrange words sequentially so that they form intelligible sentences.

Akathisia: An inability to remain still caused by an intense urge to move or fidget.

Akinesia: An inability to perform voluntary movement.

Alexia: A type of receptive aphasia resulting in the inability to read and interpret written words. Also Dyslexia: The difficulty interpreting the written word.

Alexithymia: Expressive Aphasia perceptual disorder. Inability to attach words to one's emotions; inability to express one's emotions using words. Dyslexithymia is difficulty attaching words to one's emotions.

Amygdala: An almond-shaped nucleus in the anterior temporal lobe that attaches to the caudate nucleus. May have roles in the mediation of fear and anger and the perception of social cues.

Anomie: A type of expressive aphasia that involves the inability to remember and express the names of people and objects.

Anosagnosia: Extensive neglect and failure to recognize one's own body paralysis.

Anosmia: Loss of smell (olfaction).

Anterior Commissure: Located in the anterior thalamus; allows information to travel between both thalamic lobes.

Anterior Median Fissure: Divides the medulla into equal left and right halves. The fissure continues all the way down the spinal cord.

Antitransmitter: A chemical substance that breaks down a neurotransmitter so that the postsynaptic neuron can repolarize in order to fire again. Antitransmitters terminate the postsynaptic neuron's response.

Aphagia: Aphagia means an inability to swallow. Also Dysphagia: Difficulty swallowing.

Aphasia: Impairment in the expression and/or comprehension of language. Receptive Aphasia (Wernicke's Aphasia) is the impairment in the comprehension of language. Expressive Aphasia (Broca's Aphasia) is the impairment in the expression of language.

Aphonia: An inability to make sounds. Hypophonia refers to reduced vocal force.

Aphrenia: Stoppage of thought. The individual experiences poverty of thought.

Apnea: Arrest of breathing.

Apraxia: Inability or difficulty (Dyspraxia) executing motor plans. Results from lesions to the motor cortices in the frontal lobe. Ideational Apraxia involves an inability to cognitively understand the motor demands of the task. Ideomotor Apraxia involves the loss of motor plans for specific activities; or the motor plan may be intact but the individual cannot access it.

Aprosodia: A receptive aphasia that involves difficulty comprehending tonal inflections used in conversation.

Arachnoid Mater: The middle meningeal layer located just below the subdural space. Has the appearance of a spider web.

Arachnoid Villae: The arachnoid villae are projections of the arachnoid mater into the dura mater. Cerebrospinal fluid is reabsorbed in the arachnoid villae.

Astereognosis: The inability to identify objects by touch alone. Results from damage to the secondary somatosensory area.

Asymbolia: A receptive aphasia that involves difficulty comprehending gestures and symbols.

Ataxia: Incoordinated movements resulting from cerebellar lesions.

Auditory Association Areas: Responsible for the interpretation of auditory data. There are several auditory association areas located throughout the cortex. These areas have not as yet been mapped as precisely as the visual association areas.

Autonomic Nervous System (ANS): Composed of the parasympathetic nervous system and the sympathetic nervous system. Responsible for the innervation of visceral muscles, regulates glandular secretion, and controls vegetative functions (eg, temperature, digestion, heart rate).

Axon: Fiber emerging from the axon hillock and extending to the terminal boutons. Axons transmit action potentials, or nerve signals, to the terminal boutons.

Axon Collaterals: Project from the main axon structure and serve to transmit nerve signals to several parts of the nervous system simultaneously.

Axon Hillock: The region where a neuron's cell body and axon attach.

Basal Ganglia: An unconscious motor system that mediates stereotypic or automatic motor patterns such as those involved in walking, riding a bike, and writing. Composed of three structures: caudate nucleus, putamen, and globus pallidus. Disorders of the basal ganglia often result in dystonia and dyskinesia.

Bitemporal Hemianopsia: Occurs when the temporal fields in both eyes have been lost. Results from a lesion to the central optic chiasm. Results in tunnel vision.

Blood-Brain Barrier: Consists of the meninges, the protective glial cells, and the capillary beds of the brain. Responsible for the exchange of nutrients between the central nervous system and the vascular system. Some molecules can cross the membrane while others cannot. This accounts for the inability of many pharmaceuticals to cross the blood-brain barrier.

Body Schema Perceptual Dysfunction: Body schema is a neural perception of one's body in space—formed by a synthesis of tactile, proprioceptive, and pressure sensory data about the body. Dysfunction occurs when there is a severe discrepancy between body schema and reality. Includes Finger Agnosia, Unilateral Neglect, Anosagnosia, and Extinction of Simultaneous Stimulation.

Brachial Plexus: A network of peripheral spinal nerves that supply the upper extremities. Includes C5, C6, C7, C8, and T1. Common site of compression injuries.

Bradykinesia: Slowness of voluntary movement. Seen commonly in Parkinson's disease. Also seen in depression.

Bradyphrenia: Slowness of thought. Seen commonly in Parkinson's disease and depression.

Brainstem: Composed of the midbrain, pons, and medulla. Controls vegetative functions (eg, respiration, cough and gag reflex, pupillary response, swallowing reflex).

Broca's Area: Located only in the left hemisphere, just above the lateral fissure in the premotor area. Mediates the motoric functions of speech and is responsible for the verbal expression of language.

Callosal Sulcus: Sulcus separating the corpus callosum and the cingulate gyrus.

Cauda Equina: At the end of the spinal cord—the conus medullaris—the spinal cord sends off the remaining spinal nerves that have not yet exited the vertebral column. This mass of spinal nerves is called the cauda equina because it resembles a horse's tail.

Caudate Nucleus: A basal ganglial structure involved in the planning and execution of automatic movement patterns. The caudate acts like a brake on certain motor activities. When the brake is not working, extraneous, purposeless movements appear (eg, tics, dyskinesias).

Causalgia: An intense burning pain accompanied by trophic skin changes.

Cell Body: Contains the nucleus of the neuron, which stores the genetic code of the organism.

Central Canal: Passageway through which cerebrospinal fluid flows. Begins in the caudal medulla and descends throughout the entire length of the spinal cord.

Central Nervous System (CNS): Composed of the brain and spinal cord.

Central Sulcus: Separates the primary motor cortex from the primary somatosensory cortex.

Cerebellar Peduncles: Carries sensorimotor information from the pons to the cerebellum about the body's position in space. There are three paired cerebellar peduncles: middle, inferior, and superior cerebellar peduncles.

Cerebellum: Responsible for proprioception or the unconscious awareness of the body's position in space. The cerebellum is a sensorimotor system; it receives sensory information from joint and muscle receptors concerning the body's position. The cerebellum uses this information to make decisions about how to adjust the body for the coordinated, precision control of movement and balance.

Cerebral Aqueduct: Part of the ventricular system. A narrow channel that connects the 3rd and 4th ventricles allowing cerebrospinal fluid to flow through.

Cerebral Peduncles: Large fiber bundles located on the anterior surface of the midbrain. Carries descending motor tracts from the cerebrum to the brainstem. Have an inner coat (consisting of the red nucleus and substantia nigra—collectively called the tegmentum) and an outer coat (consisting of the crus cerebri).

Cerebromedullary Cistern: (also called Cisterna Magna) Largest cistern in the subarachnoid space; allows cerebrospinal fluid to flow from the fourth ventricle to the subarachnoid space. Located between the medulla and the cerebellum. Often used as a shunt placement.

Cerebrospinal Fluid (CSF): A clear and colorless fluid that bathes and nourishes the brain and spinal cord. The composition of CSF is used for diagnostic purposes to identify disease processes.

Chemoreceptors: Sensory receptors that respond to the presence of a particular chemical; involved in olfaction and gustation.

Choroid Plexes: Vascular structures in the brain that protrude into the ventricles and produce cerebrospinal fluid.

Cingulate Gyrus: Most medial and deepest gyrus in the frontal and parietal lobes. Sits right above the corpus callosum. Shares vast connections with limbic system structures.

Cingulate Sulcus: Sulcus that separates the cingulate gyrus from other gyri in the fronto-parietal regions; located on the medial aspect of each hemisphere.

Circle of Willis: Is a circuit of five interconnecting arteries that function to prevent lack of blood flow to the brain due to occlusion.

Clasp Knife Syndrome: Involves severe rigidity at a joint. A sustained stretch will relax the muscle group and the rigidity will suddenly give way.

Claustrum: A group of nuclei located just lateral to the extreme capsule and just medial to the insula.

Clonus: An uncontrolled oscillation of a spastic muscle group that results from a quick muscle stretch. Occurs in upper motor neuron lesions.

Cogwheel Rigidity: Cogwheel rigidity is characterized by a pattern of release/resistance in a quick jerky movement. Commonly seen in Parkinson's disease.

Color Agnosia: Individuals appear to forget the concept of color. They do not appear to know the color of common objects.

Color Anomie: Individuals have lost the names for colors. However, they would still recognize that a blue banana was strange.

Commissure: Any collection of axons that connect one side of the nervous system to the other. An example is the corpus callosum.

Contractures: Limitation in joint movement due to shortening of muscles, tendons, and ligaments. Results from inactivity at a joint.

Contralateral Homonymous Hemianopsia: A loss of the visual field on the opposite side of the lesion. A left visual field cut, or left contralateral homonymous hemianopsia, results from a lesion in the right optic tract. A right visual field cut, or right contralateral homonymous hemianopsia, results from a lesion in the left optic tract.

Conus Medullaris: Is the end of the spinal cord at the L1-L2 vertebral area.

Convolutions: Are the collective name for the gyri and sulci located on the surface of the cerebral hemispheres.

Corpus Callosum: Largest commissure in the brain. Allows the right and left cerebral hemispheres to communicate with each other.

Cortex: A cortex is a layer of gray matter that contains nuclei, or nerve cell bodies. Humans have a cerebral cortex and a cerebellar cortex. The cortex sits on the surface of the cerebrum and cerebellum—underneath of which is white matter, or axons.

Corticospinal Tracts: Descending motor tracts from the primary motor cortex. Responsible for voluntary movement.

Cranial Nerves: The cranial nerves are 12 pairs of nerves that are considered to be part of the peripheral nervous system. Their nuclei are located in the brainstem and are considered to be within the central nervous system. Cranial nerves carry sensory and motor information to and from the receptors of the head, face, and neck.

Cunctation: Resisting or hindering; the opposite of Festination—which means quickened. Together, the Cunctating-Festinating Gait—characteristic of Parkinson's disease—describes difficulty initiating movement and inability to stop movement once started.

Cutaneous Receptors: Respond to pain, temperature, pressure, vibration, and discriminative touch. Found in the layers of the skin.

Decerebrate Rigidity: Damage to any tract that originates in the brainstem may result in decerebrate rigidity. Involves spastic extension of both the upper and lower extremities. The occurrence of decerebrate rigidity indicates a much poorer prognosis than does decorticate rigidity.

Decorticate Rigidity: Results from damage to the corticospinal tracts. Decorticate rigidity presents as spastic flexion of the upper extremities; spastic extension of the lower extremities.

Deep Tendon Reflexes: A reflex arc in which a muscle contracts when its tendon is percussed. Deep tendon reflexes work on the principle of the spinal reflex arc. In an upper motor neuron injury, deep tendon reflexes become hyperreflexive—because the spinal reflex arc remains intact below the lesion level causing the reflex arc to run unmodified by cortical input. In a lower motor neuron injury, deep tendon reflexes become hyporeflexive—because the reflex arc is lost.

Dendrites: The tree-like processes that attach to the cell body and receive messages from the terminal boutons of a presynaptic neuron. Dendrites can bifurcate, or produce additional dendritic branches. Bifurcation increases the neuron's receptor sites.

Dentate Ligaments: A projection of the pia mater of the spinal cord. The dentate ligaments are a series of 22 triangular bodies that anchor the spinal cord.

Depth Perception Dysfunction: Involves difficulty determining whether one object is closer to the individual than another object.

Dermatome: Skin segment innervated by a specific peripheral nerve.

Diencephalon: Collective name for the thalamus, hypothalamus, epithalamus, and subthalamus.

Dopamine (DA): The dopamine system has major effects on the motor system and on cognition and motivation. Loss of dopamine from the substantia nigra is the primary cause of Parkinson's disease. Too much dopamine has been implicated in schizophrenia. The dopamine system also plays a role in addictive behaviors.

Dorsal Horn: The dorsal horn is considered to be part of the central nervous system. It contains the cell bodies of the sensory spinal cord tracts. In the dorsal horn, the dorsal rootlets may synapse on interneurons. These interneurons then synapse with spinal cord tracts. Or the dorsal rootlets may synapse directly on the cell bodies of spinal cord tracts.

Dorsal Intermediate Sulcus: Sulci that are located just lateral to the dorsal median sulcus.

Dorsal Median Sulcus: A sulcus that divides the posterior medulla into equal left and right halves.

Dorsal Root and Rootlets: Dorsal roots are axon bundles that emerge from an ascending spinal nerve. The dorsal root leads into the dorsal rootlets—thin string-like axons that emerge from the dorsal root and synapse in the dorsal horn of the spinal cord. The dorsal root and rootlets are considered to be part of the peripheral nervous system.

Dorsal Root Ganglion: Contains the cell bodies of sensory nerves that are part of the somatic PNS. Each sensory nerve has its own dorsal root ganglion. The dorsal root emerges from the dorsal ganglia.

Down Regulation: Process by which a neurotransmitter is absorbed into the postsynaptic neuron's dendrites and cell body.

Dura Mater: The outermost meningeal layer. The dura has two projections that extend into the brain: falx cerebri and tentorium.

Dysarthria: Difficulty articulating words clearly. Slurring words.

Dysesthesia: Unpleasant sensation such as burning.

Dysmetria: Inability to judge distance. Past-pointing or over-shooting one's reach for objects. Occurs as a result from cerebellar lesions.

Dystonia: Also Dyskinesia. Both terms refer to abnormalities in muscle tone and movement. Includes Athetosis, Chorea, Parkinson's, Idiopathic Torsion Dystonia, etc

Endorphins: Work conjointly with substance P to act as pain modulators. The primary action of endorphin is the inhibition of nociceptive information.

Enteric Nervous System: An independent circuit that is loosely connected to the central nervous system but can function alone without instruction from the CNS. Located in sheaths of tissue that line the esophagus, stomach, small intestines, and colon. Composed of a network of neurons, neurotransmitters, and proteins.

External Capsule: White matter located just lateral to the putamen (of the basal ganglia) and medial to the claustrum.

Exteroceptor: A sensory receptor that is adapted for the reception of stimuli from the external world (eg, visual, auditory, tactile, olfactory, and gustatory receptors).

Extrapyramidal System: Motor structures and spinal cord tracts that do not use the pyramids to send motor messages to the skeletal muscles.

Extreme Capsule: White matter located just lateral to the claustrum and medial to the insula.

Falx Cerebri: A projection of dura mater that extends into the medial longitudinal fissure.

Figure-Ground Discrimination Dysfunction: Involves difficulty distinguishing the foreground from the background.

Filum Terminale: A projection of the pia mater of the spinal cord. The filum terminale is a slender median fibrous thread that attaches the conus medullaris to the coccyx. It anchors the end of the spinal cord to the vertebral column.

Fissure: A deep groove in the surface of the brain—deeper than a sulcus.

Flaccidity: Loss of muscle tone resulting from denervation of specific peripheral nerves. Flaccidity occurs in lower motor neuron injuries. In an upper motor neuron injury, flaccidity occurs only at the lesion level; spasticity occurs below the lesion level.

Foramen Magnum: The largest foramina in the skull—specifically the occipital bone. The opening through which the brainstem connects with the spinal cord.

Foramen of Luschka: (also called Lateral Aperture) Opening in the fourth ventricle through which cerebrospinal fluid flows into the subarachnoid space. There are two foramen of Luschka in the fourth ventricle; located in the pons. Often a site of CSF blockage.

Foramen of Magendie: (also called Median Aperture) Opening in the fourth ventricle through which cerebrospinal fluid flows into the subarachnoid space; located in the rostral medulla. Often a site of CSF blockage.

Foramen of Monro: A channel located in the ventricular system; allows cerebrospinal fluid to flow from the lateral ventricles to the third ventricle. Often a site of CSF blockage.

Form Constancy Dysfunction: Involves difficulty attending to subtle variations in form or changes in form, such as size variation of the same object.

Fornix: The fornix bodies are a pair of arch-shaped fibers that begin in the uncus and wrap around to the mammillary bodies. The fornix is a relay system for messages generated by the limbic system.

Fourth Ventricle: Region of the ventricular system through which cerebrospinal fluid flows. Located between the pons and the cerebellum. Connected to the third ventricle via the cerebral aqueduct. The fourth ventricle connects to the spinal cord via the central canal.

Frontal Lobe: Responsible for cognition, expressive language, motor planning, mathematical calculations, and working memory.

GABA (Aminobutyric Acid): A major inhibitory neurotransmitter that turns off the function of cells. GABA deficiency is implicated in anxiety disorders, insomnia, and epilepsy. GABA excess is implicated in memory loss and the inability for new learning.

Ganglia: A collection of neural cell bodies (or nuclei) usually located outside of the central nervous system.

Globus Pallidus: A basal ganglial structure involved in stereotypic or automatic movement patterns. While the caudate nucleus works like a brake on motor activity, the globus pallidus is an excitatory structure.

Glutamate (GLU): One of the major excitatory neurotransmitters of the central nervous system. Responsible for cell death when the brain experiences a major traumatic event. May also play a role in learning and memory.

Golgi Tendon Organ: Proprioceptors that are embedded in the tendons, close to the skeletal muscle insertion. Detect tension in the tendon of a contracting muscle.

Gray Matter: Sits on the surface of the cerebrum and cerebellum. Consists of nerve cell bodies.

Gyri (s., Gyrus): Are the wrinkles or folds on the surface of the cerebral hemispheres.

Hemianopsia: Field cut. See Contralateral Homonymous Hemianopsia and Bitemporal Hemianopsia.

Hemiparesis: Partial paralysis or muscular weakness of limbs on one side of the body. Occurs on the contralateral side (or opposite side) of the lesion site.

Hemiparesthesia: Loss of sensation of limbs on one side of the body. Occurs on the contralateral side (or opposite side) of the lesion site.

Hemiplegia: Complete paralysis of limbs on one side of the body. Occurs on the contralateral side (or opposite side) of the lesion site.

Hippocampus: Located within the parahippocampal gyrus. One of the major storehouses in the brain for long-term memory.

Homunculus: Cortical representation for each body part's motor and sensory function. See Motor Homunculus and Sensory Homunculus.

Hyperesthesia: An increase in sensory perception. Heightened perception of pain and temperature.

Hyperkinesia: Disorders involving speeded movement (eg, chorea, athetosis). Opposite of Akinesia. Hypokinesia is the slowing of movement just short of complete loss of movement, or Akinesia.

Hypertonia: Excessive muscle tone due to spasticity or rigidity. Opposite of Hypotonia: A lack of muscle tone.

Hypoesthesia: A decrease in sensory perception.

Hypothalamus: Two lobes (one in each hemisphere) that contain nuclei responsible for the regulation of the autonomic nervous system, release of hormones from the pituitary gland, temperature regulation, hunger and thirst, and sleep/wake cycles.

Inferior Colliculi: A pair of relay centers for audition that communicate directly with the medial geniculate nuclei of the thalamus. Located on the posterior region of the midbrain.

Inferior Olives (or Olivary Nuclei): Relay nuclei that carry ascending sensory information to the cerebellum. The sensory data pertain to the body's position in space.

Insula: A cortical region located within the lateral fissure of the temporal lobe. Site of the primary auditory area (A1).

Internal Capsule: A large fiber bundle that connects the cerebral cortex with the diencephalon. All descending motor messages from the cortex travel through the internal capsule to the thalamus, brainstem, spinal cord, and to the skeletal muscles. All sensory information travels through the internal capsule before reaching the cortex.

Interoceptors: Receive sensory information from inside the body—such as stomach pain, pinched spinal nerves, or inflammatory processes in the viscera.

Interpeduncular Fossa: The indentation between the pair of cerebral peduncles (on the anterior of the midbrain) that contains the mammillary bodies.

Interthalamic Adhesion: A commissure that connects the two thalamic lobes. Runs through the third ventricle.

Joint Receptors: Proprioceptors that are located in the connective tissue of a joint capsule. Respond to mechanical deformation occurring in the joint capsule and ligaments.

Kinesthesia: The ability to sense one's body movement in space.

Lateral Fissure: Separates the temporal lobe from the frontal lobe.

Lateral Ventricle: One of four ventricles that contains cerebrospinal fluid. There are two lateral ventricles-one in each hemisphere. The lateral ventricle is an arch shaped structure that has its anterior horn located in the frontal lobe; its body located in the parietal lobe; its posterior horn located in the occipital lobe; and its inferior horn located in the temporal lobe.

Lentiform Nucleus: Collective name for the putamen and globus pallidus (of the basal ganglia).

Limbic System: Located deep within the core of the brain, the limbic system appears to be the source of human emotions before they are modulated by the frontal lobes. The limbic system is also a storehouse for long-term memory—particularly memories that have strong emotional significance.

Locus Ceruleus: Located on the floor of the fourth ventricle in the brainstem. Secretes norepinephrine.

Lower Motor Neuron (LMN): Carries motor messages from the motor cell bodies in the ventral horn to the skeletal muscles in the periphery. A LMN is considered to be part of the peripheral nervous system. Includes the cranial nerves, peripheral spinal nerves, cauda equina, and the ventral horn.

Lumbar Plexus: A network of peripheral spinal nerves that supply the lower extremities. Includes L1, L2, L3, and L4. Common site of compression injuries.

Mammillary Bodies: Two protrusions that sit within the interpeduncular fossa on the anterior surface of the midbrain. The mammillary bodies are nuclei groups that form attachments with the hypothalamus and fornix and may play a role in the processing of emotion.

Mechanoceptors: Sensory receptors that are stimulated by mechanical deformity (eg, hair cell receptors of the skin).

Medial Longitudinal Fissure: Runs along the midsagittal plane. Separates the right and left cerebral hemispheres.

Medulla: Carries descending motor messages from the cerebrum to the spinal cord, and ascending sensory messages from the spinal cord to the cerebrum. Located between the pons and the spinal cord.

Membrane Potential: The electrical charge that travels across the cell membrane. It is the difference between the chemical composition inside and outside the cell—in other words, the sodium potassium balance inside and outside the cell. If the membrane potential is strong enough, it causes an action potential.

Meninges: The meninges are located between the skull and brain, and cover the spinal cord. They form a protective seal around the central nervous system. There are three layers of meninges: Dura Mater, Arachnoid Mater, and Pia Mater.

Metamorphosia: Involves a visual-perceptual distortion of the physical properties of objects so that objects appear bigger, smaller, or heavier than they really are.

Midbrain: The midbrain is the most rostral structure of the brainstem that sits atop of the pons and is just inferior to the thalamus. Has a role in automatic reflexive behaviors dealing with vision and audition. Site of the Reticular Activating System—has roles in wakefulness and consciousness.

Mononeuropathy: Involves damage to a single peripheral nerve; usually due to compression or entrapment.

Motor Association Area: Also called the Prefrontal Lobe. Role in the cognitive planning of movement.

Motor Homunculus: The cortical map that represents each body part's area for motor function. Location of the primary motor area (M1)—the precentral gyrus.

Multimodal Association Area: Cortical areas where sensory information merges for integration and interpretation. Two multimodal association areas: (a) Anterior Multimodal Association Area—located in prefrontal lobe, (b) Posterior Multimodal Association Area—located where the parietal, occipital, and temporal lobes converge.

Muscle Spindle: Proprioceptors that are located in the skeletal muscle. Provide a constant flow of information regarding length, tension, and load on the muscles.

Myelin: Axons are covered by a cellular sheath called myelin—an insulating substance composed of lipids and proteins. Myelin serves to conduct nerve signals. The more myelin the axon has, the faster its conduction rate.

Myotome: A group of muscles innervated by a specific single spinal nerve. The myotomes of the human body correspond closely to the dermatomes.

Neuron: The electrically excitable nerve cell and fiber of the nervous system. Composed of a cell body with a nucleus, dendrites, a main axon branch, and terminal boutons.

Neuropathy: A general term for pathology involving one or more peripheral nerves.

Neurotransmitter: Chemical stored in the terminal boutons. Released into the synaptic cleft to transmit messages to another neuron.

Nodes of Ranvier: Spaces between the myelin—on an axon—where nerve signals jump from one node to the next in the process of conduction.

Norepinephrine (NE): A neurotransmitter essential in the production of the fight/flight response, fear, and panic.

Nystagmus: Involuntary back and forth movements of the eye in a quick, jerky, oscillating fashion when the eye moves laterally or medially to either the temporal or nasal extreme visual fields.

Occipital Lobe: Responsible for the detection and interpretation of visual stimuli.

Occipital Pole: Most posterior region of the occipital lobe. Site of the Primary Visual Area (V1), responsible for the detection of visual stimuli.

Olfactory Bulb and Tract: Also known as cranial nerve 1. The olfactory tract travels directly to the hippocampus—which accounts for the deep association between specific odors and long-term memories that have emotional significance.

Oligodendrites: Compose the myelin in the central nervous system. Because oligodendrites do not produce nerve growth factor, nerve damage in the CNS cannot resolve.

Optic Chiasm: A cross-shaped connection between the optic nerves. Carries visual information from the optic nerves to the optic tracts.

Optic Nerve: Cranial Nerve 2. The optic nerves (one in each hemisphere) carry visual information from the retina, through the optic disk, to the optic chiasm. Visual information then travels from the optic chiasm to the optic tracts.

Parahippocampal Gyrus: Most medial and deepest gyrus in the temporal lobes. Folds back on itself at its anterior end to become the uncus. Relays information between the hippocampus and other cerebral areas, particularly the frontal lobes. The parahippocampal gyrus may function when humans compare a present event to an event stored in long-term memory to decide how to handle a present situation.

Parasymapthetic Nervous System: A division of the autonomic nervous system. Responsible for peristalsis and the maintenance of homeostasis.

Paresis: Partial paralysis.

Paresthesia: The occurrence of unusual feelings, such as pins and needles.

Parietal Lobe: Responsible for sensory detection and interpretation.

Peripheral Nervous System (PNS): Composed of the cranial nerves, autonomic nervous system, and the somatic nervous system.

Phantom Limb Phenomenon: The sensation that an amputated body part still remains. If the sensation is painful it is referred to as phantom pain. The cortical map of the body still retains the anatomical image of the amputated body part.

Photoreceptors: Sensory receptors that detect light on the retina of the eye.

Pia Mater: The deepest meningeal layer, located on the surface of the gyri and sulci of the brain and on the surface of the spinal cord. The pia mater of the spinal cord sends off two projections: the Filum Terminale and the Dentate Ligaments.

Pineal Gland: Innervated by the autonomic nervous system. Has a role in sexual hormonal functions and sleep-wake cycles.

Pituitary Gland: An endocrine gland that secretes hormones that regulate growth, reproductive activities, and metabolic processes.

Plexopathy: Damage to one of the plexes—brachial or lumbar. Involves multiple peripheral nerve damage.

Polyneuropathy: Involves bilateral damage to more than one peripheral nerve. An example is stocking and glove polyneuropathy. Usually caused by a disease process, such as diabetes.

Pons: Brainstem structure that acts a relay system between the spinal cord, cerebellum, and cerebrum. It largely mediates sensorimotor information on an unconscious level-shifting weight to maintain balance and making fine motor adjustments in one's muscles to perform precise coordinated limb movement. Located between the midbrain and medulla. Site of the Reticular Inhibiting System—has roles in sleep states and unconsciousness.

Position in Space Dysfunction: Involves difficulty with concepts relating to positions such as up/down, in/out, behind/in front of, and before/after.

Postcentral Gyrus: Located just posterior to the central sulcus; considered to be the Primary Somatosensory Area (SS1) where sensory information from the contralateral side of the body is detected. Also referred to as the Sensory Homunculus, the cortical representation for each body part's sensory function.

Postcentral Sulcus: Sulcus located just posterior to the postcentral gyrus.

Posterior Commissure: Located just superior to the superior colliculi. Allows information to travel between both diencephalic hemispheres.

Post-Pontine Fossa: Shallow depression located between the anterior aspects of the pons and medulla.

Postsynaptic Neuron: Second order neuron. Receives the presynaptic neuron's neurotransmitter substance from the synaptic cleft.

Post-Tetanic Potentiation: Occurs in synapses that are frequently used. When the presynaptic bouton becomes excited, it releases greater amounts of neurotransmitter substance. The postsynaptic neuron then has prolonged and repetitive discharge after firing due to too much neurotransmitter release or too slow antitransmitter work.

Precentral Gyrus: Located just anterior to the central sulcus; considered to be the Primary Motor Area (M1) where voluntary movement (on the contralateral side of the body) is initiated. Also referred to as the Motor Homunculus, the cortical map that represents each body part's area for motor function.

Precentral Sulcus: Sulcus located just anterior to the precentral gyrus.

Premotor Area: Located just anterior to the primary motor area. Has a role in motor planning or praxis.

Pre-Pontine Fossa: Shallow depression located between the anterior aspects of the midbrain and pons.

Presynaptic Neuron: First order neuron. Releases its neurotransmitter into the synaptic cleft.

Primary Auditory Cortex (A1): Located within the insula in the temporal lobe. Responsible for detecting sounds from the environment.

Primary Motor Area (M1): Location of the precentral gyrus. Where voluntary movement (on the contralateral side of the body) is initiated.

Primary Somatosensory Area (SS1): Location of the postcentral gyrus. Where sensory information from the contralateral side of the body is detected.

Primary Visual Area (V1): Responsible for the detection of visual stimuli. Located at the most posterior region of the occipital lobe.

Proprioception: The ability to sense one's body position in space. Proprioception occurs mostly on an unconscious level because it is primarily mediated by the cerebellum.

Proprioceptor: Sensory receptors located in the muscles, tendons, and joints of the body, and in the utricles, saccules, and semicircular canals of the inner ear.

Prosopagnosia: The inability to identify familiar faces because the individual cannot perceive the unique expressions of facial muscles that make each human face different from each other.

Protopathic Sensory Receptors: Adapted to identify gross bodily sensation, rather than precise regions of sensation. Detect crude touch and dull pain rather than discriminative touch and sharp pain.

Putamen: A basal ganglial structure involved in stereotypic or automatic movement patterns. While the caudate nucleus works like a brake on motor activity, the putamen is an excitatory structure.

Pyramidal Decussation: The crossing-over point where motor fibers from the left cortex cross to the right side of the spinal cord. Motor fibers from the right side of the cortex cross to the left side of the spinal cord. This is why the right cerebral hemisphere controls the left side of the body and the left cerebral hemisphere controls the right side of the body.

Pyramids: Fiber bundles that carry descending motor information from the cortex to the spinal cord. The pyramids are two large structures on the anterior region of the medulla that are divided by the anterior median fissure.

Radiculopathy: Nerve root impingement that results from a lesion affecting the dorsal or ventral roots. Can result from herniated vertebral discs.

Receptor Field: A body area that contains specific types of sensory receptor cells. Small receptor fields are located on body areas with the greatest sensitivity—lips, hands, face, sole of feet. Large receptor fields are located on body areas with less sensitivity—legs, abdomen, arms, back.

Referred Pain: Occurs when a specific body region shares its spinal nerve innervation with a separate dermatomal skin segment. The pain experienced by the body part is misinterpreted by the cortex as pain coming from a separate dermatomal skin segment.

Reticular Formation: Two systems diffusely located in the brainstem: Reticular Activating System and Reticular Inhibiting System. The activating system alerts the cortex to attend to important sensory stimuli and is involved in states of wakefulness. The inhibiting system is involved in states of unconsciousness such as sleep, stupor, or coma.

Re-Uptake Process: Process by which a neurotransmitter is reabsorbed into the presynaptic neuron's terminal boutons.

Right-Left Discrimination Dysfunction: Involves difficulty understanding and using the concepts of right and left.

Rigidity: Involves an inability to passively and/or actively move a joint on both sides of the joint. Examples are cogwheel rigidity (commonly seen in Parkinson's) and clasp knife syndrome.

Saccule: One of the receptors of equilibrium in the inner ear. Responds to changes in head position. Part of the vestibular system.

Schwann Cell: The myelin in the peripheral nervous system is composed of Schwann cells that produce nerve growth factor. This allows peripheral nerve damage to resolve (unlike damage in the central nervous system).

Secondary Somatosensory Area (SS2): Responsible for the interpretation of sensory stimuli from the contralateral side of the body.

Semicircular Canals: Receptors of equilibrium that form a system of canals called the bony labyrinth in the inner ear. There are three semicircular canals—each responds to movement of the head. Part of the vestibular system.

Sensory Homunculus: The cortical representation for each body part's sensory function. Location of the postcentral gyrus, or primary somatosensory area (SS1).

Sensory Receptor: A specialized nerve cell that is designed to respond to a specific sensory stimulus (eg, touch, pressure, pain, temperature, light, sound, position in space).

Septal Area: The region on either cerebral hemisphere that comprises the subcallosa area and the corresponding half of the septum pellucidum.

Septum Pellucidum: A sheath-like cover that extends over the medial wall of each lateral ventricle. May have a role in the processing of emotion.

Serotonin (5-HT): A neurotransmitter implicated in sleep, emotional control, pain regulation, and carbohydrate feeding behaviors (eating disorders). Low levels of serotonin are associated with depression and suicidal behavior.

Simultanognosia: Involves difficulty interpreting a visual stimulus as a whole. Patients often confabulate to compensate for what they cannot interpret visually.

Somatic Nervous System (SNS): Part of the peripheral nervous system. Responsible for the innervation of skeletal muscles.

Spasticity: Involves the inability to move a joint on one side of the joint. Usually, either the Flexors or the Extensors are spastic, but not both. Results from upper motor neuron lesions.

Spinal Reflex Arc: A spinal reflex arc is mediated at the spinal cord level; there is no cortical involvement. A sensory receptor in the peripheral nervous system sends a message along an ascending sensory spinal nerve that travels to the dorsal horn and synapses on an interneuron. The interneuron synapses with a motor cell body located in the ventral horn. The motor cell body in the ventral horn relays the message to a motor spinal nerve in the PNS, which sends the motor message to a skeletal muscle group for action in response to the initial sensory message.

Stereopsis: Depth vision.

Strabismus: Deviation of the eyeball laterally (lateral strabismus) or medially (medial strabismus).

Substance P: Acts as a neurotransmitter in the nociceptive pathway, although it is classified as a peptide. The nociceptive pathway mediates the experience of pain.

Sulci (s., Sulcus): The valleys or crevices between the gyri.

Superior Colliculi: Are a pair of relay centers for vision that communicate directly with the lateral geniculate nuclei of the thalamus. Located on the posterior region of the midbrain.

Sympathetic Nervous System: A division of the autonomic nervous system. Responsible for the fight/flight response and gearing the body up for action.

Synaptic Cleft: Space between a presynaptic neuron's terminal boutons and a postsynaptic neuron's dendrites. The terminal boutons release their neurotransmitter substances into the synaptic cleft.

Synaptic Delay: The time required for the neurotransmitter to diffuse across a postsynaptic neuron's membrane.

Synaptic Fatigue: Occurs as a result of a neurotransmitter depletion due to the repetitive simulation of a presynaptic neuron.

Tachykinesia: Speeded movement. Commonly seen in Tourette's syndrome.

Tachyphrenia: Speeded thought. Commonly seen in mania.

Tectum: The tectum is the collective name for the superior and inferior colliculi (of the midbrain).

Tegmentum: The tegmentum is the inner coat of the cerebral peduncles (of the midbrain). The tegmentum is the collective name for the substantia nigra and the red nucleus.

Temporal Lobe: Responsible for the detection and interpretation of sounds and long-term memory.

Tentorium: The projection of dura mater that extends as a horizontal shelf between the occipital lobe and the cerebellum.

Terminal Boutons: Emerge from the end branches of the axon and contain the neurotransmitter substances.

Thalamus: An egg-shaped lobe (one in each hemisphere) that contains 26 pairs of nuclei that act as a relay system for sensory and motor information traveling to and from the cortex.

Thermal Receptors: Sensory receptors that detect changes in temperature.

Third Ventricle: Part of the ventricular system. The walls of the third ventricle are created by the thalamus and hypothalamus. The lateral ventricle connects to the third ventricle via the foramen of Monro. The third ventricle connects to the fourth ventricle via the cerebral aqueduct. **Topographical Disorientation:** Involves difficulty comprehending the relationship of one location to another.

Uncus: The bulb-like anterior end of the parahippocampal gyrus.

Unilateral Neglect: Involves the inability to integrate and use perceptions from one side (the affected side) of the body or environment.

Upper Motor Neuron (UMN): An UMN carries motor messages from the primary motor cortex to (a) the cranial nerve nuclei (in the brainstem), (b) interneurons in the ventral horn. An UMN travels up to but does not actually enter the ventral horn. An UMN is considered to be part of the central nervous system.

Utricle: One of the receptors of equilibrium in the inner ear. Responds to changes in head position. Part of the vestibular system.

Ventral Horn, Root, and Rootlets: The ventral horn contains the cell bodies of the motor spinal nerves that innervate skeletal muscle. Descending motor spinal tracts travel from the cortex down to the spinal cord. In the ventral horn, the motor spinal cord tracts synapse on interneurons. These interneurons then synapse with motor spinal nerves that travel to skeletal muscle in the peripheral nervous system. The motor spinal nerve exits the ventral horn through the ventral rootlets. The ventral rootlets then merge into the ventral root. The ventral horn, rootlets, and root are all considered to be within the PNS.

Ventricular System: Hollow spaces in the brain through which cerebrospinal fluid flows. There are four ventricles: two lateral ventricles, one 3rd ventricle, and one 4th ventricle.

Vermis: The midline structure of the cerebellum that may have a role in the integration of information used by the right and left cerebellar hemispheres.

Vestibular System: Functions to maintain equilibrium and balance, maintains the head in an upright vertical position, coordinates head and eye movements, and influences muscle tone through the alpha and gamma motor neurons and the vestibulospinal tracts.

Visual Association Areas (V2 and up): Responsible for the interpretation of visual stimuli.

Wernicke's Area: Located only in the left hemisphere within the superior temporal gyrus. Responsible for the comprehension of the spoken word.

White Matter: Located beneath the gray matter in the internal regions of the cerebrum and cerebellum. Consists of myelinated fiber tracts or neuronal axons.

BUILD *Your Library*

This book and many others on numerous different topics are available from SLACK Incorporated. For further information or a copy of our latest catalog, contact us at:

Professional Book Division
SLACK Incorporated
6900 Grove Road
Thorofare, NJ 08086 USA
Telephone: 1-856-848-1000
1-800-257-8290
Fax: 1-856-853-5991
E-mail: orders@slackinc.com
www.slackbooks.com

We accept most major credit cards and checks or money orders in US dollars drawn on a US bank. Most orders are shipped within 72 hours.

Contact us for information on recent releases, forthcoming titles, and bestsellers. If you have a comment about this title or see a need for a new book, direct your correspondence to the Editorial Director at the above address.

Thank you for your interest and we hope you found this work beneficial.